What people are sayi

Paths Between Head

Oliver Robinson is a rare combination of academic rigour and courageous spirituality. We desperately need a spirituality that has a heart but doesn't require a brain bypass. This excellent book gives us a map.
Jules Evans, author of *Philosophy for Life: And Other Dangerous Situations* and *The Art of Losing Control: A Philosopher's Search for Ecstatic Experience*

As we stand at the brink of a momentous occurrence, the alignment of science and spirituality, there are few credible studies of this phenomenon. *Paths Between Head and Heart* by Oliver Robinson provides a crucial study of the split between science and spirituality and how, in fact, there is a relationship and continuum between the two. This book is an essential tool for bridging concepts in this field, even providing a way in for the skeptical who will be able to find their place in the holistic model Robinson provides.
Dr. Manjir Samanta-Laughton, author of *Punk Science* and *The Genius Groove*

This book explains the paradox of the unity of science and spirituality. Oliver Robinson shows that these two ways of looking at the world, so often seen as opposing forces, actually come from the same source, and we need to hold them in balance. Fascinating.
Jennifer Kavanagh, author of *The World is Our Cloister* and *Heart of Oneness: a little book of connection*

Oliver Robinson is an exceptionally perceptive observer of spirituality and its practices, in an age that routinely lacks insight about such matters. He is alert to both philosophical and psychological dimensions, and can bring them together. This is a book that you cannot fail to read without feeling illuminated, provoked and inspired.

Dr. Mark Vernon, author of *42: Deep Thought on Life, the Universe, and Everything*, and *After Atheism: Science, Religion, and the Meaning of Life*

This book plays a vital role in building a bridge between science and spirituality, demonstrating that they are natural partners in the evolution of our understanding of the universe and providing a comprehensive account of attempts to synthesise them. I particularly commend its dialectical structure, which puts the topic in a broader context than usual.

Bernard Carr, Professor of Mathematics and Astronomy at Queen Mary, University of London

The relationship between science and spirituality is a key issue in our time. In this brilliant, wide-ranging and engaging book, Oliver Robinson gives us a fresh take on this interface, proposing an informative new model of complementarity adapted to our age of interconnectedness. He also shows how we can integrate theory and experience, head and heart in our own lives. This is a stimulating and significant contribution to the field, which deserves to be widely read.

David Lorimer, Programme Director, Scientific and Medical Network, author of *Whole in One: The Near-death Experience and the Ethic of Interconnectedness*

Paths Between Head and Heart

Exploring the Harmonies of Science and Spirituality

Paths Between Head and Heart

Exploring the Harmonies of Science and Spirituality

Oliver C. Robinson PhD

BOOKS

Winchester, UK
Washington, USA

First published by O-Books, 2018
O-Books is an imprint of John Hunt Publishing Ltd., 3 East St., Alresford,
Hampshire SO24 9EE, UK
office1@jhpbooks.net
www.johnhuntpublishing.com

For distributor details and how to order please visit the 'Ordering' section on our website.

ISBN: 978 1 78279 900 9
978 1 78279 901 6 (ebook)
Library of Congress Control Number: 2017930560

A CIP catalogue record for this book is available from the British Library.

Design: Stuart Davies

UK: Printed and bound by CPI Group (UK) Ltd, Croydon, CR0 4YY
US: Printed and bound by Thomson-Shore, 7300 West Joy Road, Dexter, MI 48130

We operate a distinctive and ethical publishing philosophy in
all areas of our business, from our global network of authors to
production and worldwide distribution.

Contents

For my daughter Leila

A good head and a good heart are always a formidable combination.
Nelson Mandela

Acknowledgments

This book has benefitted greatly from the input of many people who have been generous with their time and wise with their feedback and suggestions. Thank you to Jennifer Kavanagh, Louisa Tomlinson and Alexander Stell for their helpful feedback on the initial book synopsis that was sent to publishers, and to Trevor Greenfield at John Hunt Publishing for his helpful communication through the process.

Thanks to Mark Vernon, Jules Evans, Jim Demetre, David Lorimer, Adrian Hornsby, Andres Fossas, Jonathan Rowson and Bernard Carr for giving their feedback on a draft of the whole manuscript. Thanks also to Nick Shannon, David Luke, Rupert Sheldrake, Eeva Kallio, Cory Barker and Charlie Morley for their feedback on specific elements of the manuscript.

Andras Gyorok, Lyn Robinson, Karina Hanson, Lucy Anker, Ilham Sebah, Nikolay Petrov and Sharath Ram Kumutaavalli have all kindly proofread draft sections at various stages of the writing process, while Elizabeth Radley provided an excellent copy-edit. Thanks to Duncan Turner for his wonderful book cover design, and to Stefania Penzes for drafting the MODI figure diagram in Chapter 10.

Having written the book mostly during evenings and weekends, and needing a space away from home to allow for concentration, I have been a regular presence hunched over my computer or a book in various cafés across South East London over the past few years. So thanks to the Valley Café in Charlton, the Costa Coffee in Next off Bugsby Way, the Star Express on Trafalgar Road in Greenwich, the Big Boyz Café in New Eltham, and the café in the Nuffield Health Gym near Chislehurst, for all being so hospitable.

Finally, my greatest thanks go to my wife Faezeh for her love and support, and for her tolerance of me and my schemes. This one seems to have come to fruition.

Preface

It is 7.30pm on a chilly December evening. My wife is out with her friends, and my 18-month-old daughter is asleep upstairs. The baby monitor is humming quietly on the living room table, and rain is tapping gently on the window. I have a few precious hours to write. Slivers of quiet time such as this, in between the demands of home life and work life, have been vital to my work on this book over the past three years. Researching it has been a gratifying and mind-stretching journey through the annals of natural science, philosophy, psychology, spirituality, and mysticism. I have run out of bookshelf space now, having accumulated and read reams of books in the process of researching this one. So it is just as well that, at the time of writing this, I am close to finishing.

My journey towards writing the book started in the late 1990s at the University of Edinburgh, where I was an undergraduate student in Psychology. In my first year, I chose to study an option course called Science and Society, mainly because it didn't have any exams. It was taught by a wise old soul called David Bloor. He asked us to reflect on the nature of scientific knowledge and to think about how it relates to philosophy, technology and religion. It was the first time I had been asked to think at this level, and I loved it. It was then that I realized my passion for connecting big ideas across disciplines, which has stayed with me all the way through the subsequent two decades, despite the pressures within my academic job towards narrow specialization.

Another important event occurred while I was at university, which acted as an early catalyst for my own spiritual journey. I went on holiday to Thailand with some friends, ready to have a few weeks of partying on the beach. After some nights out dancing and drinking, we went for a walk up to a small retreat

1

center at the top of a mountain on a small island called Koh Tau. While there, something switched on in me. I decided to leave my companions and the partying, and sign up for a retreat at the center. During a week of quiet reading, contemplation of the beautiful scenery, meditation and yoga, I realized that hiding beneath my stuffy, slightly anxious, English exterior there was a better, more awake, me.

I recall sitting on a rock jutting out of the side of the mountain, with the pastel curves of the island and the tranquil sea laid out beneath me below, and realizing in a way that was entirely beyond words that there was a sacred depth to life that my chattering mind had been hiding from me. I knew that I needed to change my life in a way that would accommodate the spiritual practices that I had started to learn, to reproduce the ineffable sense of spiritual knowing that I was gifted with at that moment. However, after returning to university I ignored the insight for several years, and got back to the business of being a student.

After leaving education, I made some conventional and ill-suited life choices. I got a job as a market research executive, and spent my days facilitating soul-destroying focus groups on the merits of power drills and sports cars. My leisure time was full of hedonism and toxins, to compensate for the meaninglessness of my career. I was caught in a rut of inauthenticity, and my emotions would not let me continue. I entered a hard and dark depression, which led to an extended period of personal crisis as I sought to find a way out of the emptiness.

I realized that I had to ditch the nice-guy persona that I had crafted to fit in, and to express my more radical side. I started to be vocally critical of society and the things that I felt were wrong with it, including what I saw as the false promises of consumerism. I wrote about this at length and got into some intense arguments with family and friends. I moved to a new part of town, started a folk band, went on a retreat in India, wrote a rather ambitious book about goals and purpose (which never got published), then traveled to Mongolia and helped to rebuild a monastery there, before starting a PhD in Psychology. In retrospect, I went through a classic 'quarter-life crisis' (something I theorize and write about now in my job as a psychologist).

Buddhism was the spiritual path I found a close affinity with at the time. I loved the way that it emphasized how everything that we take to exist as a solid 'thing' is actually formed by a web of relationships that includes the observer. I also warmed to its openness to discussing all of its ideas, as well as its emphasis on cultivating a direct experience into the nature of self and world, rather than receiving ideas on faith. I found Buddhist meditation challenging but rewarding. I occasionally glimpsed the state of luminous, pristine clarity that I had experienced all those years earlier in Thailand, and this inspired me to continue.

It was around that time that I found out about an organization called *The Scientific and Medical Network*, which explores the interface between science and spirituality via conferences, lectures, and publications. I loved the big topics of the events such as time, consciousness, dreams and purpose, and the way that the members were willing to express unconventional, original thoughts without fear of censorship. I became a director of the organization for some years and still do voluntary work for them part-time. As I passed out of my twenties and into my thirties, I found my own spiritual inclinations moving away from Buddhism towards Quakerism, and became a regular attender at Quaker meetings. Quakerism defies all stereotypes of religion, having no priests, churches, theology or hierarchy. It is, like Buddhism, deeply experiential and mystical.

After concluding my PhD, I gained work as a lecturer in Psychology, and I have been an academic for just over a decade now. My specialism is the field of adult development. I research how humans shift, transform and grow during adulthood, particularly through major transitions and crises. I have written a textbook on this topic called *Development through Adulthood*. I also write about psychological research methods, and on how philosophy and psychology intersect.

Psychology is a boundary place, aspiring to be a hard science but humming with the conundrums of subjectivity and free will, both of which science has traditionally preferred to avoid or deny in its pursuit of objective fact. Some psychologists work hard to make the discipline more like the natural sciences, while others pull in the other direction to make it more focused on uniqueness, subjectivity and the unpredictability of the individual. Following the philosopher-psychologist William James, who produced work that fits with both of these paradigms, I try to integrate them in my work and show how they are complementary.

The idea behind this book germinated in 2011, when I gave talks in London, Berkshire, and Bath that outlined the basic concept in early form. In 2013 I put together a book proposal and submitted it to publishers. Since getting the publishing contract, it has taken three years to work the book up into a final form, in between the demands of a busy job and starting a family.

Researching the book has changed me in a number of positive ways. My passion for natural science, which was particularly strong when I was young, has been re-ignited thanks to all the reading I have done. I have developed a practice of meditation and currently do it every day. I have found a passion for spiritual practices that involve the body, particularly the moving meditation practices of Five Rhythms (5Rhythms) and Biodanza. I have also trained in shamanic journeying, which has revealed to me how consciousness in certain altered states opens into realms of extraordinary transcendental otherness. I still do shamanic retreats and workshops on occasion. I also attend Unitarian services from time to time, and find that Unitarianism offers an inclusive and warm community that encourages spiritual development, and honors difference while providing a common language of spiritual values.

The structure of the book is based on a seven-part scheme

for showing how science and spirituality are harmonious partners in the quest for truth and wholeness. The scheme is novel in its scope and structure, despite it resting on many pre-existing foundations. Chapters 1 and 2 set the scene, Chapters 3 to 9 gradually build up the argument, and then in Chapter 10, I present an integrative model. In the Epilogue, I discuss why I think an integration of science and spirituality matters for our time. We are, I will argue, at a crucial junction in the human journey and we need to bring all our faculties together to manage a cultural shift into a radically interconnected world. Science and spirituality are important partners in this.

So it's time to start. I hope you find the book to be interesting and illuminating, and look forward to discussing it with you at some point in the future, should our paths cross.

Chapter 1

Setting the Scene

When Albert Einstein was in his late fifties, he wrote an article called *Physics and Reality* in which he concluded that science rests on unknowable mysteries. Why does the universe show an elegant and beautiful form? Why does it seem to have a hidden structure that is comprehensible with conceptual theory and algebraic equations? From where or what do scientific laws come, and why do they seem to govern the physical world? These questions had brought Einstein to a conviction that the theories and formulae of scientific knowledge convey only part of reality. Beyond them, he surmised, lie the immeasurable, the inexplicable and even the miraculous.[1]

You too have mysterious layers that science struggles to reach. Consciousness, subjectivity, meaning, purpose, and morality defy the objective lens of the scientist in part or whole. I will never know what it feels like to be you, nor you me, no matter how much science we learn. Furthermore, many of the deepest experiences in your life, such as the feeling of unconditional love or the sublime beauty of a natural scene, are impossible to fully describe in the words and numbers upon which science depends. They require more subtle modes of depiction. The areas that science struggles to reach and explain are the natural territory of spirituality – it thrives in the unknown and transcendental, and in the subjective depths of lived experience.

Science and spirituality have both separated from religion over the course of the modern era into different, yet complementary, domains of inquiry. Spirituality still draws on ideas and practices from organized religion, while taking a more experiential and eclectic ethos to its subject matter than the latter. In many ways, it can be seen as religion's unconventional and inquisitive younger sibling. To understand the formative give-and-take between spirituality and religion is an important first step towards making sense of the rest of the book, so it is to this topic that I turn first.

Comparing spirituality and religion

Organized religion combines sacred rituals, rules, beliefs, texts, and codes of behavior into a formally recognized social institution. The first religion that included a written scripture appeared in Ancient Egypt around 3000 BCE. Over the subsequent millennia, organized religion spread across the rest of the world, as civilizations grasped the varied benefits of structuring the spiritual impulse into manageable groups and hierarchies.[2] Indeed the word hierarchy has religious origins – it comes from the Ancient Greek word *hierarkhes*, which means sacred rank.

The great religions of Buddhism, Judaism, Christianity, and Islam, all of which emerged between 500 BCE and 500 CE, perfected this system of formalizing and containing the spiritual life within circumscribed institutions. They all developed an explicit membership system, whereby a convert would state their allegiance, publicly become a member, and henceforth agree to adopt the conventions and rules of the collective. Membership of one religion precluded being a member of another, which meant clear boundaries were established between religious communities. In return for this loyalty and singular commitment to the group, members were promised a defined path to salvation or enlightenment.

Today, the major religions continue to use this group membership system. To become a member one must undergo a joining or conversion ritual, then adhere to the core beliefs and practices that are required for membership. In drawing together large numbers of individuals under an agreed set of conventions and a common purpose, religious groups are powerful structures indeed, and this power has historically been used for good and ill. The charitable and educational work of religious groups has been enduring and widespread, but the corruption of religion for violent or controlling ends has also been immense. Weighing these positive and negative effects against one another is hard, for they are not directly comparable in a quantifiable sense. Supporters of religion will tend to focus on the positive side, critics on the negative side, and arguments continue to this day.[3]

Over the past four centuries, religion has been challenged by a more recent kind of experimental, de-institutionalized approach to matters of the sacred. Since the early twentieth century, it has been mainly referred to as spirituality. It includes the open exploration of topics such as ultimate purpose, transcendence, the divine, spiritual healing, yoga, meditation, states of consciousness, enlightenment, sacredness, prayer, love, ecstasy and the nature of the soul or higher self.[4] In contrast to religion's emphasis on social stability and continuity, spirituality emphasizes exploration, transformation and growth.[5,6] Correspondingly, at the center of religion one finds a strong focus on the past, in the shape of historical scripture and the upholding of traditions. In contrast, at the core of spirituality one finds a focus on the future, in ideas and practices that pertain to the realization of higher potentials in self and society.[7,8]

In comparison to the communal framework provided by religion, the personalized approach of spirituality leads to more pluralism of belief, with individuals often drawing on multiple sources of inspiration.[9] Although this may seem like a loose 'pick-n-mix' approach, it has been used productively by philosophy for centuries. Students and experts in philosophy find their own integration from the many theories and arguments in the field. It is neither expected nor desired that all philosophers should think the same, or come to the same conclusions. Indeed it is precisely the dynamic differences between them that keep philosophical ideas as living truths rather than dead dogma. This same kind of eclecticism invigorates spirituality, and keeps its ethos helpfully distinct from religion.

While many choose to pursue spirituality without affiliating to a religion, they are not mutually exclusive and can be combined in productive ways.[10] Many liberal religious groups now accept the value of reaching out spiritually into areas beyond their own boundaries and practices. Notable examples are the Quakers, Unitarians, Liberal Anglicans, liberal strands of Sufism such as *The Sufi Order*; liberal Judaism exemplified by the Rabbi Michael Lerner's *Network of Spiritual Progressives*; as well as many moderate strands of Buddhism, Jainism and Hinduism. These open and tolerant religious groups in the West tend to be less visible and vocal than the more exclusivist and fundamentalist sects, so the atheists and agnostics who look at religion from afar may well only hear the loud shouts of the hard-liners, and may miss the subtle and open messages that are proffered by religious moderates and liberals.

A negative manifestation of how religion and spirituality can mix is the phenomenon of the modern cult. Many cults are based on the same transformative practices and ideas that are explored openly in spirituality, but within a highly controlled environment that has fixed boundaries, a lack of questioning, and deference to authoritarian leaders. Cults are manifestations of what happens when spirituality is led away from free exploration and autonomous thinking towards dangerous credulity. I will argue over the course of the book that it is precisely the input of the scientific mindset that prevents spirituality from drifting into gullibility and cultish forms.

The harmonies of science and spirituality

Over the course of the book, I will be putting forward the case that science and spirituality are harmonious partners in the quest for truth, well-being and wisdom. Harmony, whether in ideas, music or art, occurs when different things that contrast with each other are combined to create a higher-order whole. The word harmony comes from the Latin *harmonia*, which means joining together. In music, harmony is created when different notes with complementary frequencies come together in such a way that they combine into a higher unity. In art, harmonious combinations are produced by juxtaposing colors that are opposites on the 'color wheel' (such as orange and blue, or green and red). In the domain of ideas, harmony is created when two or more contrasting ideas bring a greater understanding together than one does alone, despite the tension between them. In sum, all kinds of harmony involve a paradoxical mix of both difference and commonality.

In considering science and spirituality, certain key commonalities provide the foundation for harmony. Firstly, both science and spirituality are transformative and active ways of knowing through experience. One can read about them and talk about them, but that is unlikely to lead to any progress in either. *Doing* them is achieved by undertaking particular kinds of embodied activity over an extended period, after receiving the right kind of training. For the scientist, the physical activity that is necessary is data collection from the external world, via traveling to the data collection site, gathering field notes, taking measurements, making observations and specimen-gathering. For the spiritual practitioner, the embodied practices used to facilitate development include meditation, yoga, tai chi, centering prayer, dance, singing and playing music, psychotherapy, psychedelic exploration, ethical activism, and helping activities such as charitable work or caring for the sick.

By pursuing the right methods, and by accepting a fair amount of trial and error, the assumption in both science and spirituality is that a practitioner will develop a more accurate conception of reality, and so move closer to the truth and further from falsity.[11]

Science pursues truth through its methodology that links the collection of external evidence with mathematics and reasoned thinking. The intention in the scientific method is to elicit dispassionate and objective knowledge about the external world, which transcends any individual point of view and is superior to common sense.

Spirituality pursues truth not as something beyond subjective consciousness but as a state of awakeness and higher awareness within it. Through practice, the seeker connects with a "ground of being" beyond ego, which is felt to be a source of authentic love, compassion and peace, that connects people and other living things together.[12] It is as though we are all cups of ocean water, and through spiritual practice, we eventually realize our true identity as ocean, not cup.

A second key commonality is that both science and spirituality entail reflective questioning, criticality, and a wariness of dogma. Within science, critical thinking is highly valued and all research is scrutinized by other scientists as part of the peer-review process, to help ensure quality control.[13] Scientists are encouraged to self-criticize too, and to constantly reflect on limitations and ways of improving their methods and theories.

The reflective and critical processes of spirituality are more informal than those of science, but no less important. Mature spiritual questioning entails reflecting on whether what is being experienced or learned via one's practice is congruent with reason and intuition, and helpful to personal and social development.[14] Critical reflection is further facilitated by talking to others and by perusing the ever-expanding literature on spirituality. Feelings are integral to the reflection process, as often they are gut indications about issues that conscious thought sometimes struggles to grasp.[15]

In addition to these commonalities of (a) learning through experience and (b) critical reflection, the harmony of science and spirituality is a product of their complementary differences, seven of which I look at in this book. I define these seven using the following pairs of opposites:

1. Outer   Inner
2. Impersonal   Personal
3. Thinking   Feeling
4. Empirical   Transcendental
5. Mechanism   Purpose
6. Verbal   Ineffable
7. Explanation   Contemplation

Between each pair of opposites there runs a spectrum of difference that creates a 'path' between the concept shown on the left, which is traditionally associated with science, and the concept on the right, which is associated with spirituality. This is a matter of degree rather than absolute; the methods and activities of science have an *emphasis* on the left-hand side concepts, while spirituality has an *emphasis* on the concepts listed on the right-hand side. Within each spectrum of difference, there are intermediary positions that combine and hybridize science and spirituality, like shades of gray created by mixing black and white.

In Chapters 3 to 9 of the book, I journey down each path in turn, stopping off at various points from one end to the other, including frontier areas that sit at the interface between science and spirituality. In Chapter 10 I draw the threads of the discussion together, by proposing how the polarities can all be seen as expressions of one fundamental duality.

Seven is an auspicious number for the scheme, given that it is rich in scientific and spiritual meaning. On the scientific side, it has unique mathematical qualities; for example, it is the only number between 1 and 10 that cannot be divided or multiplied to make a number between 1 and 10. It is found in a variety of natural systems, including seven colors in a rainbow, seven notes in both major and minor musical scales, seven rings around

Saturn, and seven diatomic elements in the periodic table.

In religion and spirituality, seven also has a special place. In Christianity, Islam and Judaism, there are seven days of creation, and Christianity refers to seven deadly sins and seven virtues. The Book of Revelation is full of the number – there are seven churches, seven spirits, seven seals and seven trumpets. In the Talmud and Koran there are seven heavens, and when Muslims pilgrimage to Mecca they walk around the Kaaba seven times to represent this. Seven is prevalent in Eastern religion too; in Hinduism there are seven higher worlds and seven lower worlds, while the system of subtle energies that underpins yoga conceives of seven chakras in the human body. Given all this scientific and spiritual resonance, it is no surprise that in a recent multinational poll, seven was found to be the world's favorite number.[16]

Those of you who are familiar with the science-religion debate will notice passing similarities between the scheme presented in this book and the *Non-Overlapping Magisteria* (NOMA) theory of science and religion developed by the paleontologist Steven J. Gould.[17] Like Gould, I suggest that science has limitations that allow space for other ways of knowing, but there are fundamental differences between his scheme and mine. Firstly, Gould's scheme only refers to Christianity. In contrast, this book is about spirituality and hence not about any one religion in particular.

Another key difference is that Gould proposes that science and religion cannot mix. He uses the analogy of oil and water to visualize this – if oil and water are put in a jar, they create two distinct layers and don't mix even at the join. That, says Gould, is how religion and science are. In contrast, I propose that science and spirituality definitely *do* mix and overlap. There are scientific approaches to spirituality and spiritual approaches to science, and this interface area between the two is a fascinating and controversial area that I look at in various ways across the chapters of this book.

The seven dialectics all contribute to an integrative model, which is presented in the final chapter. I call it the *Multiple Overlapping Dialectics* (MODI) model. But let's not get ahead of ourselves – there's much groundwork to do first in presenting all seven elements of the model. This allows the big picture to emerge gradually and digestibly. The next task is to briefly outline the process of how to think with dialectics and polarities, and how that relates to positive human development.

Dialectical thinking

Dialectical thinking has been employed for millennia by philosophers, theologians, mystics, scientists, and psychologists as a way of reconciling the tension of opposites, and of integrating apparent contradictions. It is often referred to as 'both-and' thinking due to its preference for seeking to integrate two contrasting ideas, rather than choosing one and rejecting the other. Perhaps the most famous advocate of dialectical reasoning in the West was the philosopher Hegel. He proposed that ideas evolve through a process of dialectical interaction, whereby an idea or theory (a thesis) is presented, and then opposing or critical ideas are presented (an antithesis). As a dialogue between the two sides proceeds, resolution occurs through a higher-order synthesis that integrates both the thesis and antithesis. This synthesis of the two then becomes a new thesis, and the process starts all over again. Over time, this process of dialectical challenge and integration leads to greater harmony and oneness in knowledge.

Dialectical thinking is also exemplified in Chinese yin-yang philosophy. The yin-yang conception of reality and health has been at the center of Chinese culture for thousands of years. It conceives of two principles – yin and yang – which manifest in everything as two complementary properties. Yang is associated with externality, objects, thinking, hardness, and solidity. Yin is associated with internality, feeling, softness and hiddenness. They are opposite but complementary, separate but interpenetrating, and constantly in flux. To find wholeness and health in life involves finding a balance of yin and yang. An excess of either causes illness or dysfunction in a person or culture. Following this dialectical tradition of balancing opposites, Chinese culture is traditionally skeptical of all extremes, preferring the balanced middle ground.[18]

Dialectical thinking entails seeing how every idea is inextricably linked to its opposite. For example, a dialectical approach to well-being sees happiness as inextricably linked to its opposite, sadness, and sees good fortune as linked inevitably to misfortune. Thinking about well-being in this way allows for a certain acceptance of life's ebb and flow. The following ancient Chinese story captures this ethos:

> Once upon a time, there was an old man on the frontier. One day he lost his horse. This was a very bad thing for him. But very soon, the lost horse came back home and brought another horse with it. This was very good. Then his son got injured when he rode on one of these two horses. So, good fortune brought misfortune. But, later on, young men were called to go to the army and join a war. Since his son was injured and he was not able to be a soldier, his son was safe. So misfortune again brought the old man good fortune.[19]

In this story, each fortunate event paves the way for a misfortunate one, and vice versa. So what appears at any one moment as fortune or misfortune is a mixture of both when seen from a longer-term focus. The two, being opposites, are forever intertwined. The story also shows how dialectical thinking tends to emphasize the formative role of the observer. What is seen as fortune or misfortune depends on the frame of the reference of the observing individual. For the dialectician, *all* opposites show this same interdependence on each other and an observing mind.

One can further illustrate this with the geographical concepts of east and west. East and west only exist relative to an observer (for example, the USA is to the west of Europe, but to the east of Russia), and relative to each other (you can't have the idea of east without the idea of west).

Dialectical thinking is different to the formal logic that comes from Aristotle, which underpins much thinking and debate in the West. In formal logic, one of the three key 'laws of thought' is called the law of the excluded middle. This law entails an *either-or* mindset; it states that *either* a statement is true, *or* its opposite is true. It doesn't allow for the possibility that two contrasting ideas may both contain a grain of truth. It thus tends towards a simplistic notion of truth and falsehood, in which some people are completely right, and others are completely wrong. This is sadly prevalent in religion and politics, leading to partisan positions and difficulty in engaging in productive dialogue.

Fritjof Capra, in his book *The Tao of Physics*, discussed how dialectical thinking has been important to science since the arrival of quantum theory in the early twentieth century.[20] The early quantum physicists needed to resolve deep paradoxes in the behavior of subatomic particles and waves, and they found inspiration from the East, in the dialectical thinking style of Taoism and yin-yang philosophy. An example paradox that needed a *both-and* solution to be solved was the nature of light. They found that light appears to scientific instruments as *both* particle *and* wave, depending on how it is observed and measured. They called this complementarity and realized that it is beyond the understanding of standard *either-or* logic, for waves and particles are structured completely differently and have different effects on measuring devices.

In summary, the dialectical thinker considers contrasting ideas and aims to synthesize them where possible, while honoring the differences and tensions between them. So when considering juxtaposed ideas, such as the differences between science and spirituality, a dialectical approach seeks a solution that includes both within a higher harmony, but that does not attempt to reduce one to the other.[21]

Dialectical thinking and human development

Dialectical thinking is linked to human development in several important ways.

Carl Jung, whose ideas I return to periodically over the course of the book, proposed that personality is a tug-of-war between competing tendencies, including extraversion-introversion, sensing-intuition, and masculine-feminine. To develop one's personality positively is to find a balance between these polarities. This is done by engaging in activities that enhance the underdeveloped side of each one. Because we are always learning new ideas and experiencing new sensations, any balance that is achieved is always temporary. In the process of trying to regain balance, we move forwards in development towards a state of greater wholeness and unity. This process, called *individuation*, is never-ending.

Another influential theorist of human development who placed the idea of dialectical balance at the heart of his theory was Erik Erikson. Erikson's theory conceives of stages of development across the lifespan, each of which has a core dialectic that the person must balance, before moving on to the next stage. For example, in adolescence it is *identity vs. confusion*, in early adulthood, it is *independence vs. commitment*, and in later life, it is *integrity vs. despair*. If a dialectical balance between these opposites is not found at any stage, then a crisis ensues during which a person seeks urgent solutions to their imbalance.[22]

A further way that dialectical thinking relates to human development is that the capacity for dialectical reasoning has been found to typically appear in adulthood, following the development of formal logic in the teenage years. Michael Basseches, in his book *Dialectical Thinking and Adult Development*, concludes that dialectical thinking stems from an increased capacity in adults to question the boundaries around a problem, and to combine different perspectives in order to find higher integrated harmonies in knowledge and living.[23,24]

Robert Kegan's theory of development also states that dialectical thinking is indicative of stages only reached during adulthood. His model posits sequential levels of mental complexity, and characterizes the Self-Transforming level as one that is built around dialectical thinking.[25] At this level, a person sees all ideas and categories in dynamic mutually-forming relationships, centering on the interactive relationship between subject and object.

The dialectical approach to understanding science and spirituality, as well as being accurate in a factual sense, also represents an aspiration for development such that the logic and rigor of the head, and the feelings and intuitions of the heart, find a healthy balance. I will argue in Chapter 10 that science and spirituality represent these two archetypal facets of the human mind and brain, and that to explore them in tandem provides for wisdom. There are many other ways of developing your head and heart, but for those who are interested in science and spirituality, they both provide rich storehouses for the task.

Within chapters 3 to 9, I provide practical exercises for you to engage in that will help to develop your scientific and spiritual awareness in ways that relate to the concepts presented. To develop your scientific side is to enhance your knowledge and awareness of the world around you, to take an impersonal, unbiased view of yourself and others, to link rigorous and methodical observations to reasoned predictions and theories, and to understand the mechanisms and mathematics that describe how things work. To develop your spiritual side is to explore your own inner life through meditative and/or ecstatic practice, to connect with other people compassionately and kindly as beings worthy of your care and attention, to explore expanded levels of awareness, to experience more love and less fear, and to explore life's purpose and meaning on your own terms. These two trajectories of development work particularly well in combination, for being opposites they correct the tendency in the other towards excess.[26]

Chapter 2

Entangled Histories

Science and spirituality have both grown exponentially over the past four centuries. Science's rise to prominence has been well documented, spirituality's has been less so, and their historical parallels have been rarely considered at all. When one does look at their modern histories together, a clear pattern emerges. It becomes apparent that revolutionary periods in science have coincided with related revolutions in spirituality, and that common features underlie these parallel revolutions across four ages, which I call the *age of rebellion, the age of wonder, the age of paradox*, and the *age of interaction*. Understanding this entangled historical picture lays important foundations for the dialectical arguments presented in subsequent chapters, many of which dip back into this history in parts.

The age of rebellion: The scientific revolution and the non-conformist revolution

The first parallel upsurge of science and spirituality occurred in a dark period in Europe's history. The time in question was the mid-1600s. The apocalyptic Thirty Years War (1618–1648) had left a quarter of the population dead in some European countries, and the bloody English Civil War (1642–1651), followed by the interregnum period of 10 years, had thrown the established feudal system into doubt. The divine right of the monarchy, and its guiding assumption of a fixed hierarchical order in the universe, would never recover its medieval luster. Social changes were creating new crises too. Urbanization was increasing rapidly, and illnesses such as cholera and the plague, which spread easily in the dirty and crowded cities, were rife. Life expectancy in large British cities was under forty, as low as it had been since the Dark Ages.[1] People across the continent were hungry for new ideas and solutions.

Growing literacy and reduced printing censorship were having an effect on society too. Wider reading meant more people dared to ask whether the old ways of strict hierarchical religion and deference to tradition had run their course and needed replacing. Radical groups in religion, philosophy and politics sprang up, all of which had the same revolutionary and optimistic notion at their heart. This idea was simply that a better future was possible if, and only if, individuals were free to question tradition and to innovate new ideas without fear of reprisal. The medieval ethos to which these modern radicals were opposed was not so enamored with the future. It viewed human culture as being in a decrepit state compared with the high points of the ancient past.[2] Traditions originating in the old times were seen as ways of maintaining a connection with past glories and perfections, and so should be upheld exactly.

The rediscovery of the philosophy and art of Ancient Greece during the Renaissance had challenged many aspects of the religiously-structured medieval worldview, but Renaissance scholars were still in thrall to the past. They considered the philosophers of Ancient Greece to be representatives of a higher ancient race of humans, whose ideas could be debated but not improved on.[3] Hence academia in the early 1600s centered principally on interpreting the works of Aristotle and Plato, and studying the geometry of Euclid. In the words of historian Rupert Foster Jones:

Dazzled by the recovered light of the past, the Elizabethans so invested the ancients with the robes of authority that the latter became oracles, to question which bordered on sacrilege. They restricted the study of nature to the narrow confines of a library… They faced backward rather than forward in the quest for truth, and prepared themselves for scientific investigation by mastering the Greek and Latin languages. The natural result was an ardent worship of the great classical minds, and a submission to their authority which is hard for us to appreciate.[4]

Theories and concepts from ancient philosophers were debated extensively, but the notion that new theories could be developed from scratch was rarely considered, even when an old theory did not fit with the evidence. For example, the theory of the four bodily humors, developed originally by Hippocrates (460–376 BCE), was still used as the basis for medicine in the seventeenth century, two thousand years after it had been developed. The theory proposed that health problems, both mental and physical, were the result of an imbalance of four fluids – black bile, yellow bile, phlegm, and blood. The theory was patently false as no one had ever seen black bile, but in this and other matters, people still stuck with theories from the revered ancient past rather than follow the evidence of their senses.

It was in the late 1600s, just after the aforementioned time of general crisis in Europe, that a growing confidence arose that new systems of understanding could be developed that would be superior to the ancient ways. Science gained its first official recognition at this time, and the scientific revolution took off after slow beginnings. The *Académie des Sciences* was established in 1665 in Paris, and *The Royal Society* was founded in London in 1663 to support the study of natural philosophy and science. The motto of the Royal Society is *Nullius in Verba*, which means 'take nobody's word for it.' It encapsulated the rebellious and anti-authoritarian spirit of the age. Royal Society members were asked to doubt all received wisdom and instead seek new truths through methods of observation and quantification. The successes of Isaac Newton during the 1680s in devising a new quantitative system of natural laws were fuel to this fire; they showed that one individual, with sufficient hard work and inspiration, really *could* improve on the ancient past.

Other scientific breakthroughs of the time added to the enthusiasm for new knowledge. Galileo discovered the craters of our moon and that there were moons circulating Jupiter; Kepler discovered that the planets move around the sun in elliptical orbits; Robert Hooke discovered the biological cell; Nicholas Steno developed a theory that fossils were organic remains of past species; William Harvey discovered that the heart operates blood circulation; Christiaan Huygens developed a theory of oscillating motion and designed the first pendulum clock; and Ole Romer measured the speed of light.

While science was becoming independent from philosophy and theology during this period, religious groups were at the very same time breaking away from the Church to explore the spiritual life on new terms, also motivated by a rebellious spirit of questioning. These societies, such as the Seekers, Baptists, Muggletonians, Ranters, and Quakers, were spiritual in focus but were not recognized as religions at that time. This meant an increasing ambiguity over what was and was not religious activity. In 1662, they were collectively labeled 'non-conformist sects' by an Act of Parliament. The term non-conformist unintentionally provided an ideal badge of recognition for their intentionally rebellious ethos, and they willfully adopted it. Science and these new dissenting religions had much in common – they both spoke of progress and human betterment, of a world with greater rights for individuals, of more openness to innovation and change, and more trust in experience and reason.

The Quaker movement (the official name for which is *The Religious Society of Friends for Truth*) was founded by George Fox in the 1640s, and it continues to exist to this day as a radical offshoot of Christianity. The Quakers have no priests, no holy communion that involves wine and bread, no set theology, and no churches. They believe that religion is internal to people's hearts and souls, rather than an external membership to a group.

They argue that people who have never read the scriptures can be spiritual if they live by the promptings of love and truth in their hearts. The arrival of the Quakers was an early modern inkling that spirituality beyond religion was on its way, and it is no accident that many influential early Quakers were women, at a time when all priests were men.

The seventeenth century also saw the rise of mystical religious movements in continental Europe, including the Pietists in Spain, the Behmenists who followed the mystic Jacob Boehme, and the Rosicrucians in Germany and England. Like the Quakers, these movements emphasized *experience* as a path to spiritual truth, over and above the authority of scripture and faith. The Rosicrucians espoused an early version of modern 'syncretism', through creatively combining elements of Christianity, Hermetic philosophy, Kabbalah, and Gnosticism. Like the Quakers, they stated that unmediated experience of God is possible for those who are willing to devote themselves to particular meditative and transformative practices.

All of these new spiritual movements blurred the distinction between the sacred and the profane. The spiritual impulse had started to leak out of the church and into the secular world, and there would be no stemming of this tide. The trickle would become a flood, and eventually, de-institutionalized spirituality would end up defining modernity as much as the rise of science.

The age of wonder: The chemical revolution and the romantic revolution

The scientific revolution of the seventeenth century had been fueled by developments in mechanics and physics. Kepler, Newton and others had shown that motions on Earth and in the heavens follow quantitative laws that can be conveyed in equations. Despite these advances, the nature of *what* was in motion had yet to be understood. A hundred years after Newton's *Principia* had been published, scientists still conceived of matter as comprised of the four elements of air, water, earth and fire. Some scientists had suggested adding a fifth element of 'aether', while a more adventurous fix had been attempted by Georg Stahl in 1703, who renamed the fire element 'phlogiston.' According to Stahl's theory, combustion was thought to be the release of phlogiston from whatever was being burnt.

Chemistry was not only being held back by defunct theories, but also by limitations with the apparatus needed for rigorous experiments with substances, liquids and gases. Chemistry experiments generally require three things: a reliable source of heat, a device to measure heat, and sensitive scales to measure changes in weight. These instruments were all developed in the 1700s: oil lamps were devised that produced a constant heat source, mercury thermometers became widely available in the 1720s, and shortly after, Anders Celsius devised a fixed scale for measuring heat. In 1770, the first set of spring scales was devised by Richard Salter, which revolutionized the capacity to detect small changes in mass.

These new pieces of apparatus meant that quantitative chemistry was possible by the late 1700s, and this new approach to the discipline was first pioneered by Scottish chemist Joseph Black and English vicar Joseph Priestley. Yet it fell to a French chemist Antoine Lavoisier to revolutionize chemistry by throwing out the theory of four elements and phlogiston, and developing a new conceptual foundation for the discipline. Aided by his wife Marie, he developed a theory of combustion that relied on a new element, which he named oxygen. He also named hydrogen, phosphorus and fifty-two other elements that together comprised a new table of basic elements. Other innovations of his included ways of labeling compounds using suffixes such as 'ide' and 'ate', e.g. copper sulfide and copper sulfate. He set this new scheme out in *Traité Elementaire de Chimie*, which was published in 1789. Tragically, Lavoisier fell victim to the French Revolution and was guillotined for tax improprieties in 1794.

The chemical revolution of Lavoisier and Priestley not only provided chemistry with a solid scientific foundation, but also transformed science more generally by making it far more explosive, colorful and dramatic than the quiet and diligent work of earlier physicists and mathematicians. The reactions of chemistry experiments often involved fire, explosions, and dramatic visible transformations of one substance to another. This gave science a new level of public appeal. On the back of the new understanding of heat and gases that chemistry had provided, hot air balloons were invented (the first was launched in 1783), providing the general public with further evidence of the adventurous, boundary-breaking nature of science.

While science became more swashbuckling and explosive, spirituality also became more adventurous and fun. In contrast to the rather anti-frills revolution of the earlier non-conformist movements, the romantic revolution brought art and romance to modern spirituality. It did so in ways that were a major challenge to the religious institutions of the time.[5] Romanticism emerged towards the end of the 1700s out of a diffuse collective of artists, poets, musicians and philosophers who sought to pursue spiritual truth through art, dreams and contemplation. As science pursued general laws and uniform order, Romanticism moved in the opposite direction by seeking creative ways of expressing uniqueness and particularity. Richard Tarnas, in his book *The Passion of the Western Mind*, describes this as follows:

In contrast to the scientist's quest for general laws defining a single objective reality, the Romantic gloried in the unbounded multiplicity of realities pressing in on his subjective awareness, and in the complex uniqueness of each object, event, and experience presented to his soul.[6]

The importance of being unique and authentic was championed

in human character too. Eccentricity was seen by the romantic as a sign of a person who was thinking for himself or herself, and hence waking up to the self-reflective life, rather than simply following the majority or unthinkingly adhering to social rules. This, in turn, fueled a new interest in libertarian philosophy that emphasized individual autonomy, freedom of choice and the respect of differences.[7]

The great composers of the Romantic era were seen by the general public as new nonreligious saints, and they were often explicit about the spiritual aspirations of their craft.[8] Public music concerts became a new kind of congregation where people could have a transcendental experience beyond religion.[9] Religious music also moved out of churches, which caused controversy. When Handel's *Messiah* was first performed at a concert hall in Dublin instead of a church, bishops and clerics protested bitterly at this mixing of sacred and profane. Romantic art and poetry also moved away from religious iconography towards a mystical or animist portrayal of nature. Romantic music dwelt on nature-based themes, such as Vivaldi's *Four Seasons*, Beethoven's *Moonlight Sonata*, Tchaikovsky's *Waltz of the Flowers*, Debussy's *Clair de Lune* (Moonlight), and Chopin's *Raindrops*.

For Romanticism, the inner dimensions of consciousness were considered a wellspring of truth, beauty and creative power. Dreams, visions, altered states and creative inspiration were considered of divine origin, and intuition and the emotions of bliss and wonder were seen as more direct means to great truths than hard rationality.[10] Creativity was held aloft as a human capacity that most closely reflected Nature's own capacity for creation. God for the Romantic was an artist, not a mathematician, and the universe was a testament to His love of beauty and elegance.[11]

As the works of the great Romantic poets, composers and artists flooded into public awareness, it became apparent to the general population that beautiful music and art, irrespective of whether it had a religious topic or not, could induce spiritual experience. Furthermore, contemplating Nature was increasingly seen not as a pagan sin but as a portal to the divine.

This period that gave birth to both Romanticism and the chemical revolution has been referred to by historian Richard Holmes as the *Age of Wonder*. Holmes argues that the same sense of wonder, adventure and curiosity underpinned both Romanticism and science during this period. They overlapped in 'romantic science', as represented by the pioneering chemistry of Humphry Davy, who not only discovered sodium, potassium and chlorine, but also personally explored mystical states of consciousness using a recent product of chemistry – nitrous oxide.[12] This mixing of mysticism and science was to become more important during the subsequent age of paradox.

The age of paradox: The relativity-quantum revolution and the mysticism revolution

By the late nineteenth century, science seemed to rest on firm foundations. The universe appeared to be made of solid matter that worked in predictable and mechanical ways. But any celebrations that science had achieved a final and reliable picture of the cosmos were premature. The edifice of predictability came crumbling down in the early twentieth century, as the discoveries of relativity theory and quantum theory inserted paradox and uncertainty into the fabric of reality, in ways that are still challenging scientists now.

Einstein's relativity theory showed that space and time are not absolute. They seem fixed and certain to us, but that is just our human perspective. They in fact only exist relative to observers, and to frames of reference that observers impute. They are part of a four-dimensional *spacetime continuum*, which dilates and contracts in relation to the speed of the observer. So the faster the observer is moving, the slower time passes and the more space contracts. For example, astronauts on the International Space

Station, who are orbiting the Earth at 17,000 miles an hour, age slightly more slowly than people on the surface of the planet. At the speed of light, time disappears and space contracts to zero, so a photon traveling from a distant star will have taken many years to make the passage from our perspective, but will not have aged even slightly. To look at it another way, if one were to catch a ride on a beam of light, the whole universe and its history would be present in a mysterious singularity. There would be no distance between things, and no time elapsed between the beginning of the universe and now.

At the same time that relativity theory was leading to new insights and conundrums, quantum theories were being developed to make sense of the behavior of subatomic particles such as electrons and photons. The quantum revolution was led by pioneering scientists such as Erwin Schrodinger, Niels Bohr and Wolfgang Pauli, who were enthusiastic explorers of Eastern mystical philosophy and used ideas from mysticism to help make sense of the strangeness of the subatomic world. It was discovered that electrons and photons accorded to an *uncertainty principle* that meant one could only talk about them in terms of probabilities. Sometimes they behaved like waves, spread out in space and time, and then change into particles when observed (more on that in Chapter 5). Electrons were found to be particularly strange. They teleport from one orbit to another in an atom, without moving through any intermediary positions – this is called a quantum leap. They do not exist in one location, but rather within a cloud of probability. Because they don't have one fixed location, if you place a barrier next to an electron that it cannot physically cross, it occasionally appears on the other side. This effect is called quantum tunneling and has been shown in many experiments.[13] One can predict that it will happen from time to time, but never exactly when. It is predictably unpredictable.

After the rise of relativity theory and the quantum revolution, science was on a very different footing from before. Gone was the universe of determined predictability and billiard-ball causality that Newtonian physics seemed to show. The cosmos was now conceived as having paradox at its most fundamental level. During this same historical era, spirituality went through a parallel revolution based on uncertainty, relativity and paradox. This was in large part due to the rise in popularity and awareness of mysticism.

A mystic pursues a first-person experience of the divine or the absolute, but what he or she finds is inexpressible in words or precise formulations, hence one must accept uncertainty and imprecision in any third-person spiritual formula or religious theology. The Eastern traditions of Buddhism and Hinduism, both of which have strong mystical elements, flooded into the public awareness for the first time towards the end of the nineteenth century and the beginning of the twentieth.[14] New spiritual movements that emerged during the time, including Theosophy (founded in 1875) and Anthroposophy (founded in 1912), incorporated ideas and mystical practices from Buddhism and Hinduism.

In 1893, representatives of the world's many faiths met in Chicago, at the Parliament of the World's Religions. This was the first time that interfaith dialogue on such a scale had been achieved, and with it came a notable increase in the awareness of Eastern spiritual traditions in the West.[15] A central basis that made this interfaith dialogue possible was that the attendees at the event assumed a common *mystical* core to all valid religions, which existed behind and beyond the different theologies. This idea of a mystical basis for all authentic religions was set out by the philosopher William James in *The Varieties of Religious Experience* (1901), and by Evelyn Underhill, an English poet and novelist, who wrote *Mysticism: A Study of the Nature and Development of Man's Spiritual Consciousness* (1911).

Popular art movements of the time such as Impressionism

and Symbolism incorporated this new cultural fascination with ambiguity and uncertainty into their visual imagery. Impressionism rests on the idea of presenting images of natural scenes that are hazy and imprecise and hence provide an element of uncertainty onto which the viewer may project their own interpretation. An influential post-impressionist group called the Nabis (a Hebrew word meaning 'prophets') proposed that by depicting ambiguous forms and feelings within art one could convey spiritual truths that were paradoxical and multivalent.[16]

The growing interest in mysticism also led to a new movement that focused on the health benefits of meditation and spiritual practice, called New Thought. It stated that human beings could become more mentally and physically healthy through daily spiritual devotion and practice. Books by well-known New Thought writers sold in the millions, for example, Ralph Waldo Trine's *In Tune with the Infinite* (1897) sold over 1.5 million copies, and Ella Wheeler Wilcox's *The Heart of the New Thought* (1902) was a hit with American and European audiences alike. New Thought presented an optimistic alternative to the sin-laden confessional practices of the churches. It was far from just inward-looking solipsism – it was associated with progressive social action movements of the time, including the emerging feminist movement.[17]

During the age of paradox, the paranormal became an increasing focus of interest in academia and the general public. This was partly influenced by the rise in interest in Hindu and Buddhist mysticism, which state that mystics develop paranormal powers or 'siddhis', such as telepathy or telling the future. The Society for Psychical Research was founded in 1882 to investigate such phenomena from a nonreligious perspective, and its work continues to this day.

Importantly for this book, the word spirituality was given its contemporary meaning during the age of paradox. The first time it was employed in print to refer to something that is separable from religion was in a little book written by Felix Adler in 1905 called *The Essentials of Spirituality*. Adler was an American-Jewish academic and social reformer who had been strongly influenced by Kant and Emerson. He wrote that "spirituality is not indissolubly associated with any one type of religion or philosophy; it is a quality of soul manifesting itself in a variety of activities and beliefs."[18] This was a re-branding of an old word for new times.

The age of interaction: Postmodernism, the dark revolution and the spirituality revolution

Since the 1970s, there has been an unprecedented period of fertile interaction between science and spirituality. This increase in contact between the two is a measurable fact. For example, the medical journal search engine *Medline* shows that in the 1980s just 30 peer-reviewed academic papers were written in medical journals with the word spirituality in the title. In the 1990s this number went up to 247. Between 2000 and 2009, the number leaped up to 1063. From 2010 to February 2018 (the time of writing), the number was at 1102. In recent years, scientific research on the efficacy of prayer and meditation has become a major growth area across various scientific disciplines (more on that in Chapter 3). Meanwhile, ways of being scientific and evidence-based about spirituality have been increasing in popularity and prevalence.[19]

Various factors are likely to be at play in driving this increasing interaction. Firstly, the rise of postmodernism in the Seventies provided more scope than ever before for mixing different paradigms and frames of reference, including spiritual ones. Postmodernism emerged in academia and art as a challenge to science and rational schemes that purported to explain reality in one way, and instead proposed that reality is in large part a product of a person's perspective, and hence there are many ways of conceiving reality and representing it. This pluralism that defines postmodernism was by no means new – it has a strong heritage in the libertarian strand of modernity that called for different points of view to be voiced and heard (for example the work of John Stuart Mill). However, in the 1960s and 1970s, postmodernism took pluralism to a new level, deconstructing the grand narratives of science and religion, and providing new spaces for exploration of spirituality and alternative means of inquiry outside of established paradigms.

Another factor behind the increasingly open dialogue between science and spirituality has been the increasing realization in physics over recent decades about how much of the universe we *don't* understand. The discoveries of dark matter and dark energy are examples of this; their discovery has been said to constitute a new scientific revolution; the dark revolution.[20] Dark matter is a hypothetical substance, the gravity of which accounts for how light moves through space and how galaxies stay together. It is thought to account for 23 percent of the mass of the universe. Dark energy is used to explain how it is that the universe is continuing to expand, and why this expansion is speeding up rather than slowing down. Energy has an equivalent mass (a fact that is at the heart of relativity theory), and hence dark energy is thought to comprise 73 percent of the mass of the universe. This means that all the visible matter in the universe – all the stars, planets and interstellar dust – comprises just 4 percent of the universe. So physicists, with all their best scientific instruments, can access only a small fraction of the cosmos. The rest – up to 96 percent of the total mass – is composed of substances and energies that as yet can't be observed and don't have a clear theoretical explanation with classical or quantum physics.[21] This has provided powerful evidence that current science does not provide a complete picture of reality, and so may leave room for other ways of knowing.

Another reason for the increased interaction between science and spirituality is that the latter has become a lot more popular in recent decades. Philosopher David Tacey refers to this recent sudden rise as the *spirituality revolution*.[22] Research supports this social shift, showing a radical increase in the popularity, accessibility, and variety of spiritual literature and practices outside of traditional religion.[23] Spiritual experiences have become markedly more common over the same period. In 1962, a Gallup poll found only 22 percent of Americans reported having a spiritual experience. In 1976 that number had gone up to 31 percent, and then in 1994, a poll by *Newsweek* on the topic reported 33 percent. Then in 2006 an IPSOS poll found that 47 percent of people reported a spiritual or mystical experience, and a Pew Forum poll in 2009 saw that number increase again to 49 percent.[24]

To handle the growth of nonreligious spirituality, demographers have coined the name 'spiritual but not religious' (SBNR) to categorize the portion of the population who define themselves as such. Major panel surveys suggest up to 30 percent of Europeans and 25 percent of Americans now refer to themselves in this way.[25,26,27] So we are no longer dealing with a countercultural fringe; these figures suggest a higher prevalence of SBNR than atheism and most religions in the UK.

The rapid expansion of spirituality has meant that commercial interests have moved in, keen to maximize its profit-making potential. With that has come a host of products for sale to spiritual seekers, quick-fix solutions for happiness or inner peace, and popular shops that sell crystals, angel cards, wind chimes, fairy statues, incense and relaxation music. This high street 'New Age' version of spirituality has become its most public manifestation but is no more than a pastel-toned parody of what I shall be discussing in this book.

The field of transpersonal psychology has also been an important forum for the meeting of science and spirituality. It was founded in the mid-1960s to further the theoretical study of spiritual experiences and spiritual development. I discuss it more in Chapter 6. A number of interdisciplinary organizations have spearheaded science-spirituality dialogue over recent decades, drawing on transpersonal psychology in the process. They include the *Scientific and Medical Network*, the *California Institute for Integral Studies*, and the *Institute of Noetic Sciences*.[28]

In summary, the contemporary spirituality that pervades Western culture is a varied and subtle movement that has four hundred years of heritage in philosophy, religion, psychology, mysticism, art, and music. It has developed in close parallel with science through this time, intertwining and entangling but remaining distinct. One key figure who drew both of these modern strands into his work was Carl Jung, and we turn to his ideas now, as context to the first of the seven polarities – outer and inner.

Chapter 3

Outer – Inner

Carl Jung was a reclusive character, preferring the quiet life of the country to the buzz of the city. The readers of his works knew very little about him until, at the age of 81, he unexpectedly decided to write an autobiography called *Memories, Dreams, Reflections*.[1] The book focuses in large part on the inner spiritual experiences and crises that shaped Jung's life. He describes how in his childhood he had a strong sense of destiny accompanied by frequent and intense religious visions. Both the visions and the sense of destiny were kept hidden from others, for he worried that divulging them would lead to ostracism or ridicule. As a teenager, he continued to read up on spirituality and philosophy in private, while becoming fascinated by science. When deciding what to study at university, his family encouraged him to follow his scientific interests, and he went to the University of Basel to study for a degree in science and medicine.

At university, the spiritual and scientific aspects of Jung's character had become two distinct subpersonalities, which he referred to as "Number 1" and "Number 2." Number 1 was his outer public personality – it was characterized by a zest for scientific knowledge, fossil hunting, reading scientific journals, and studying. Number 2 represented Jung's inner spiritual side – this side of him felt a keen inner closeness with the divine, was fascinated with the deeper meanings of dreams and visions, and was drawn towards mystical and esoteric ideas. These two sides to Jung's character finally started to reconcile and combine when he started studying psychoanalysis. He described being at his first lecture on Freud as a moment of epiphany – like a bridge being lowered down to join his two sides together.

Jung conceived of spirituality and science as focused on the inner and outer world respectively. The form of psychoanalysis that he developed would be a means of linking these, and as a result, its nature was neither pure science nor spirituality but a hybrid of the two. For many who were looking for a way of connecting science and spirituality in their own life, Jungian psychoanalysis was an exciting new path to follow.

As a key part of his broad-ranging theory, Jung proposed that the direction of all human development is towards a state of wholeness, by balancing opposites. No doubt influenced by his own sense of being split into an inner self and outer self in his early life, he proposed that a central polarity for balance in development is between *introversion* and *extraversion*. Introversion is defined by attention and energy towards the inner life of the mind, while extraversion is the tendency to look outwards at physical and material phenomena. People, and cultures as a whole, tend to favor one or the other of these, but health and sanity are found in balancing the two. Following Jung, I argue in this chapter that science is modernity's extravert impulse to explore the outer world, and spirituality is its introverted impulse to attend to the depths of inner life.

Science and the search for external evidence: Primary and secondary qualities

A common feature of all branches of science is that evidence in support of a theory or law must be drawn from the outside world. It is not enough to put forward a solid argument or logical proof as evidence in science, for this is not considered sufficient guard against speculation. A researcher must observe, measure, record and catalog objects and events in the outer world in some way, bring back data and show how ideas and theories fit with that.

There is a fundamental initial challenge with exploring the outer world, which is that it is difficult to ascertain what is genuinely 'outer' in human experience, and what is not. This is because in human experience, what appears to be the world beyond the body is really a mix of outer information from the senses and inner additions from our minds. Take a green leaf for example; we see the leaf as an object in the external world, but the perception of it as green is something that the mind adds to it – the color green is *not* out there. Hence people who are color-blind will see a different color, and animals who don't have color receptors in their eyes (such as bats and seals) will see no color at all. Due to an intricate process of cognitive blending, we see the greenness of the leaf as external to us with the rest of the leaf. Inner and outer aspects thus merge into one experience, and it can be quite a challenge teasing them apart to work out what is actually external and what has been added in the act of observation.

Ever since Galileo, science has a standard solution to this problem. It distinguishes two kinds of qualities that objects have – *primary* and *secondary* qualities. Primary qualities are quantitative – they include shape, motion, number, figure, and size. Science asserts that primary qualities are 'out there' in the world, as properties of objects themselves, and so are present irrespective of who is observing and even if no one is. Secondary qualities, in contrast, are those that are imputed by the mind of the observer. They include smell, taste, color, sound, significance, and meaning. They are products of the observing subject, not the observed object, so the theory goes. Galileo wrote about secondary qualities as follows:

I think that tastes, odors, colors, and so on are no more than mere names so far as the object in which we locate them are concerned, and that they reside in consciousness. Hence if the living creature were removed, all these qualities would be wiped away and annihilated.[2]

Science typically assumes secondary qualities to be effects of primary ones. For example, the color green in the leaf is explained as the effect of the quantitative properties of the wavelength of light traveling from the leaf to the retina. Secondary properties are therefore assumed to be less fundamental than primary properties. They are real insofar as they are part of lived experience, but they are not *really* part of the physical world with its objects, forces and quantitative properties.

The primary-secondary quality distinction, popular and enduring as it is, is questionable. For one thing, primary properties are all mathematical, and hence are represented via symbols and abstract measurement scales that have been developed by human beings. They require the observer to attach these symbols and scales to the object. Given this, primary qualities arguably require an observing mind just as much as secondary ones, despite being more reliable and easy to agree on than the latter. This line of reasoning was proposed by the philosopher Kant, who argued that reality 'as it is' (what he called the *noumenon*) does not have any qualities, whether primary or secondary. Its nature is a mystery and both primary and secondary qualities are features of human experience.

This idea of Kant's has recently been resurrected within the theory of biocentrism, postulated by scientist Robert Lanza.[3] Lanza draws on the theory of relativity and quantum physics to argue that all definable qualities of objects, and the objects themselves, exist in *observed* reality. For example, motion is a key primary quality, but the theory of relativity has shown there is no absolute motion. Motion is always *relative* to a frame of reference. As a simple example, imagine a jeep going at 60 miles an hour, and a person fires an arrow from a crossbow directly out of the back of the jeep at precisely 60 miles an hour. It would stand still in the air relative to the ground, but would move away from the car at high speed. So relative to the car, the arrow is moving, but relative to the ground it is still. In other words, the arrow is *both moving and not moving*, depending on your frame of reference. If there is no frame of reference imputed by an observer, there is no way of ascertaining if anything is moving at all. This gets clearer at larger scales of reality – is the Milky Way moving? Yes, but only relative to other galaxies. If the Milky Way was the only galaxy in the universe, we would have no way of knowing if it was moving or not, and indeed it wouldn't make sense to refer to its movement without other galaxies to provide a comparison point.

If Kant and Lanza are right, then the primary-secondary quality split as a basis for distinguishing inner from outer is tenuous. Yet, whether philosophically sound or not, distinguishing primary and secondary qualities has given science a basis for focusing on the mathematical features of nature that can be measured reliably by observers and can form testable laws and models. This has led to quantitative theories that are highly replicable and testable, and to a wealth of evidence and fact about the world that is generally agreed across cultures and individuals. Gathering this abundance of evidence about the physical world has proved to be quite an adventure for science, for it has entailed all kinds of explorations to the many corners of the world and solar system, in order to bring home the data.

Scientific exploration: Reaching ever further outward

The extraverted ethos of science is clearly expressed in its commitment to exploring the outer world. Over the centuries, scientists have organized expeditions across the surface of the Earth, the seas, the sky and into space, constantly seeking to push the boundaries of our knowledge of the cosmos. Every environment that can be reached has been cataloged, mapped, and theorized about.

The first great scientific expeditions were across the sea. As the great European voyages of discovery set sail to chart the territories of the Earth, scientists were often on board. They returned with evidence of new species, materials, peoples, and lands. They showed that the external world was larger, more exotic, more diverse and more investigable than anyone had previously imagined. The material world, so often maligned by medieval religion, was in fact magisterial. One of the most famous of the pioneering explorer-scientists was Sir Joseph Banks (1743–1820), who took part in James Cook's voyages between 1768 and 1779 to Newfoundland, Brazil, Tahiti, New Zealand and Australia. Banks brought hundreds of species to the attention of European scientists, including penguins and marsupials, as well as many trees and plants. Eighty species now bear his name.

In the nineteenth century, maritime expeditions by Charles Darwin and Alfred Russel Wallace were crucial to the development of the theory of natural selection. Both wrote up their expedition experiences as books, which became bestsellers and helped to cement the legendary status of scientific exploration as a key means for furthering our understanding of nature.[4]

Earth still holds many secrets, and expeditions across its surface continue to this day. A major frontier for science is now the deep sea, 95 percent of which remains unexplored.[5] The Amazon rainforest is also far from exhaustively researched, and as of 2015, there exist an estimated 67 uncontacted human tribes in the rainforests of Brazil alone, who live in the distant reaches of the rainforest in isolation from the outside world.[6] Barely anything is known about them. It is up to scientists to find out, but the danger with contacting such tribes is that contact with other human groups can kill, as they lack immunity to many common diseases. Thus scientists are currently studying the isolated tribes from the air, using high-magnification cameras.[7]

As well as scientific expeditions across the surface of the world, expeditions *upwards* have been central to the thrust of science's exploration of the outer world and its endeavor to conquer the heavens with reason and evidence. The pioneering scientific voyages upwards came with the first hot air balloons in the late 1700s. The balloons vindicated science's early theories of gases and their response to heat, while adding exciting new views of the planet's surface for cartographers and geographers. Over the centuries, exploration upwards has progressed further and further from the planet's surface. Unmanned space probes are the great explorers of our time, providing evidence from across our solar system of the substance and structure of planets, moons and asteroids.

Not only does scientific exploration bring increased awareness of the external world, but it changes how we view ourselves. The early maritime scientific expeditions showed modern man how much more there was in the world by way of species, landscapes and food. Space probes have also changed our perspective on ourselves. On 14th February 1990, the space probe Voyager 1 turned its camera back towards Earth and took a photo of our planet from a distance of 6 billion kilometers, famously entitled *Pale Blue Dot*. Earth's size within the image is less than a pixel. The scientist Carl Sagan wrote of this photograph:

> The aggregate of our joy and suffering, thousands of confident religions, ideologies, and economic doctrines, every hunter and forager, every hero and coward, every creator and destroyer of civilization, every king and peasant, every young couple in love, every mother and father, hopeful child, inventor and explorer, every teacher of morals, every corrupt politician, every 'superstar,' every 'supreme leader,' every saint and sinner in the history of our species lived there – on a mote of dust suspended in a sunbeam.[8]

The extraverted thrust of scientific exploration has gradually diminished the apparent importance of human beings, as it has revealed that the cosmos is large beyond imagining, and that we are tiny in the overall scale of the universe. However, in relation to the Earth, it has given human beings the power to reshape nature in radical ways via external technologies. It is to the technological face of science's extraverted agenda that we turn next.

Exercise for developing outer awareness: Connecting with your solar system

Thanks to science, far more is known about the solar system than ever before. We know what all the planets look like and what they are made of. Yet the majority of people are unable to recognize the planets in the night sky, and with that their sense of connection with the solar system is lacking. Thanks to technology, this modern deficit is now easy to remedy.

This exercise is to learn to recognize the five planets of our solar system that are visible to the naked eye at night – Mercury and Venus, which orbit closer to the sun than the Earth; Mars, Jupiter, and Saturn orbiting further away from the sun than the Earth. First, download the app *Star Chart* for your smartphone or tablet. Go into display settings and turn off constellation images. Then with the app switched on, point the device at the sky and it will show you what planets and stars you are looking at. Learn the positions of the planets relative to your location at different times of the evening and night. Get a sense of their path (planet paths are shown as dotted lines on the screen). You will see that all planets move on the same circular track through the night sky, as they are all on the same plane around the sun. See how you feel when you can eventually spontaneously look at the sky and recognize the planets – you have added a new level of outer awareness to your life.

The app will also show you where the sun is beneath the horizon at night. If you stand outside at night, with your screen orientated so that you can see several planets above you, and the location of the sun beneath you, you may suddenly have a direct experience of the Earth in orbit with the other planets around the sun, on the backdrop of the starry canopy. To add to this sense of being on a traveling planet, try the following exercise proposed by Brian Swimme in his book *The Hidden Heart of the Cosmos.*[9] Go outside at night, lie down on the grass, and try to perceive that you are underneath the Earth, peering *down* into the chasm of the night sky, while being pulled upwards into the planet through gravity. Achieving this kind of perceptual reorganization is empowering and fun, and helps as a reminder that you are stuck to the edge of a sphere hurtling through space, with no up or down beyond that which your mind imputes.

The science-technology relationship: A key to investigating the outer world

Science's mastery of the external world is driven not only by explorations but also by its close relationship with technology. Over the course of its history, science has driven the development of new technologies and then used many of these within its research to aid data collection or analysis. Some of the earliest examples of this two-way link between science and technology were the telescope and the microscope. These were developed from the science of optics and then subsequently were used by scientists to make key discoveries. The same can be said of the thermometer, the barometer and countless more instruments that emerged as applications of scientific theory, and then became instruments in research.

A more recent example of this formative interaction between science and technology is the use of microwave detection devices in research into the origins of the universe. Microwaves are a type of electromagnetic wave that has a shorter wavelength than radio waves but longer than infrared. They were first predicted by James Clerk Maxwell in 1864 from his theory of electromagnetic radiation, and then Heinrich Hertz proved their existence in 1888 by building a device that produced and detected microwave radiation.

Having created microwave detection machines, science has subsequently used them to test the Big Bang theory of the universe's origins. In the 1960s, physicists were still debating whether the Big Bang theory or 'steady state theory' best accounted for the universe. The former said that the universe had a defined beginning, but the latter said it did not. One of the predictions of the Big Bang theory was that there should be a microwave radiation residue from the original explosion of the Big Bang, which would still be rippling through the cosmos now, and hence detectable from space (but not from the surface of the Earth). In 2001, a microwave detection satellite called the WMAP was launched, and the data it sent back to Earth fitted exactly with the predictions of the Big Bang theory. It found cosmic microwave background radiation. Thus Maxwell's original theory of electromagnetic radiation had, a century after it was devised, led to a machine that had helped to further cosmology.

The overall effect of science and technology working together has been an extensive reshaping of the biosphere. This impact of science is visible from outer space, for example in power-generating structures such as the Three Gorges Dam in China (which is twice the height of Big Ben and over a mile long), and open-pit mines such as Bingham Canyon Mine near Salt Lake City (which is 3 miles across and 1 mile deep). Also perceptible from space is that now two-fifths of the world's land area has been turned over to agriculture. Particularly visible are the large, factory-style farms that cover thousands of hectares and rely heavily on technology and science.

These examples of the human capacity to alter the physical world represent what geologists call the Anthropocene epoch (Anthropocene from Greek, meaning 'the age of man'). This is the period of Earth's history, starting in the twentieth century, during which the combined activities of human beings became such a defining force in shaping Earth's atmosphere, oceans and landmasses that their effects appear in rock strata. As humans have co-opted land for intensive farming and industry, many other ecosystems have been destroyed in huge quantity. Over 20% of the Amazon rainforest has been lost to human activity, and rainforest continues to be lost at approximately 80 million acres a year.

Tragically, evidence suggests that we are in the midst of a mass extinction of species that is resulting directly from the human impact on the biosphere.[10] Ninety-nine percent of threatened species are at risk due to human activities that have led to habitat loss, the introduction of alien species into ecosystems, and global warming. According to some shocking estimates, 30 percent of species could be extinct within a century from now.[11]

This level of destruction would be impossible without the power that science has granted human beings. Nevertheless, scientists are now trying hard to resolve the environmental destruction and projected future damage that their disciplines have been complicit in creating. Sustainable and environmental technologies are pouring forth from laboratories, including high-tech methods of recycling, renewable energy sources such as solar power and wind power, methods for the purification of polluted air and water, biodegradable plastics, devices that promote energy conservation, and technologies that can be used to combat global warming.

In summary, the outer focus of science has changed our understanding of, and relationship with, the world around us, in ways that have broadened awareness but strained the biosphere. Science's external focus has also reshaped our understanding of ourselves, as researchers have sought whether all inner life is explainable through outer data and theories. It is to this story – the endeavor to scientifically externalize the life of the mind – that I briefly turn now, prior to discussing the inner path of spirituality.

The science of mind: Turning inner into outer

The science of psychology emerged historically when a number of pioneering scientists set out to discover whether the inner phenomena of mental life could be translated into external data. New methods were needed for this, and over three decades (1870–1900) there was a flurry of innovation. Here I briefly describe four such methods that have had enduring influence in psychology: observation of behavior, psychometric tests, neuropsychological measurement and qualitative methods.

One of the early pioneers of observation in psychology was none other than Charles Darwin, who after his success with the theory of natural selection wrote *The Expression of the Emotions in Man and Animals* in 1872. Darwin aimed to show that the intangible concept of emotion could be brought within the corpus of scientific theory through the observation of facial expressions. Following Darwin's example, the school of behaviorism focused specifically on observing behavior as a source of psychological data. Early behaviorists were highly motivated to align their work with the science's focus on external objects, and some researchers even attempted to remove any mention of inner life at all.[12]

Other forms of observational research do infer inner feelings and thoughts. For example, one of the most well-known observational methods is used to measure the emotional attachment of young children to their caregivers. It is called the 'strange situation experiment.' In this 20-minute assessment, observations of the child are made when the mother leaves the room, and when a stranger tries to interact with him or her.

From the behavioral observations, the researcher will infer whether there is an emotional bond. This inference of an inner *felt* phenomenon from outer behavior allows the scientist to stay focused on measurable externals, but requires an inference from those externals to something that is subjective.[13]

Psychometric methods employ *tests* and *questionnaires* to turn the intangible phenomena of attitudes, traits, emotions, and self-concepts into external measures of quantity. The first intelligence tests were developed by Francis Galton in the 1880s. Psychometric instruments became particularly popular in the 1940s, and they now play a central role in contemporary psychology, for they are now the most widely used way to quantify inner experiences into outer data.

Over recent decades, the development of brain scanning methods such as Magnetic Resonance Imaging (MRI) and EEG (Electroencephalogram) has brought ways of linking inner phenomena to brain activity. In research that uses brain scans, participants will conduct a psychometric or cognitive test, or engage in some kind of structured activity, and then undergo a brain scan. Alternatively, individuals with a mental illness diagnosis or learning disability will be scanned and compared with a group of normal controls. This form of psychology attracts a high level of scientific prestige not only due to its external and physical focus, but also its formative link with cutting-edge technology. For its followers, it holds out the promise of a purely external and object-based study of mind.

Not all methods in psychology aim for reducing the mind to numbers or brain activity. Qualitative methods externalize experiences by capturing them in written words. For example, a person may be asked to describe their experiences in an interview, and the transcription of the interview becomes the external data that science requires. Many qualitative methods aim to convey the subjective experience of research participants in detail. In so doing they acknowledge the reality of the inner life of human beings more than other methods, but they do so at the expense of some traditionally accepted criteria of science, such as replicability and detachment. I return to this topic in more detail towards the end of Chapter 9.

In attempting to bridge the inner and outer world, psychology frequently finds itself the target of criticism from both the natural sciences and from spirituality. Natural scientists question whether psychological theories and research findings are genuinely replicable in the way that findings in physics, chemistry and biology are. This is currently a major topic of debate, following a recent project that attempted to replicate major psychological studies and failed on most counts.[14] Yet human beings react to science in different ways from other animals and objects – knowing a theory about oneself can actually change one's behavior, which can mean the theory may not apply the same way the next time. For example, learning the theory of goal-setting, which was developed by Locke and Latham in the 1980s, may likely change how you set your goals.[15] Given that people change in response to being theorized about, psychology is unlikely to ever have the same replicability and objective status as the natural sciences. That is no bad thing – the relative unpredictability of human beings is to be celebrated.

The spiritual approach to first-person inner experience is different from scientific psychology. The aim in spirituality is not to study other people and try to turn their experiences into external data, but to explore one's *own* mind at the first-person level, journeying into consciousness to encounter its deeper layers and its luminous source. The spiritual path has notable parallels with the process of undergoing psychoanalysis, which also involves intensive first-person introspection, and we will return to ideas from Jungian psychoanalysis later in the chapter to help make sense of things.

Spirituality and the journey within – historical roots

In contrast to the appetite of science for all things external, even in the study of mind, spirituality has a more introverted thrust to many of its activities, with a fascination for the depths of the inner life and the first-person mode of inquiry. The historical roots of this are the mystical traditions upon which contemporary spirituality has drawn. The mystical sects of religion have traditionally taken an inward and meditative approach to the spiritual life. A Christian theological dictionary definition written in 1856 describes mystics in a way that emphasizes this:

> MYSTICS (n)... Under this name some comprehend all those who profess to know that they are inwardly taught of God. The system of the mystics proceeded upon the known doctrine of the Platonic school... that the divine nature was diffused through all human souls... they maintained that silence, tranquility, repose, and solitude, accompanied with such acts as might tend to attenuate and exhaust the body, were the means by which the hidden and internal word was excited to produce its latent virtues, and to instruct men in the knowledge of divine things... They lay little or no stress on the outward ceremonies and ordinances of religion, but dwell chiefly upon the inward operations of the mind.[16]

As implied in the above definition, most forms of mysticism are less ritualistic and dogmatic than their mainstream counterparts due to their emphasis on the *experience* of the divine, over and above theology or the adoption of religious conventions. For example, the Quakers, a mystical sect of Christianity, seek the "inner light" of consciousness as the manifestation of the divine within, and pay attention to the promptings of love and light in the heart as spiritual messages. Quakers are relatively uninterested in theology or ritual, having abolished most of Christianity's conventions, including baptism and the use of priests.

Scriptures tend to be seen by mystics as non-literal signposts to help on the inner path rather than literal accounts of outer events. For example, in Sufism, the term *jihad* principally refers to 'inner jihad' – the struggle against fear and anger within, rather than struggles against enemies without. In mystical Christianity, Heaven and Hell represent states of mind dominated by love and fear respectively, rather than actual places. In the words of John Milton, "the mind is its own place, and in itself can make a Heaven of Hell, and a Hell of Heaven."[17]

Mystics employ methods of meditation, contemplation and silent retreat to cultivate a still mind. Over time these practices are said to bring about an inward realization that consciousness shares a sacred bond or identity with the divine or the absolute. This mystical precept has also become the central pillar of modern spirituality. One of the most famous early voices for spirituality beyond religion was TW Higginson, whose essay "The Sympathy of Religions" was a manifesto of the Parliament of the World's Religions in 1893. Higginson believed that the outer trappings of religion were ultimately relative, for the real spiritual authority lies within. He wrote:

The soul needs some other support also; it must find this within; – in the cultivation of the Inward Light; in personal experiences of Religion; in the life of God in the human soul... In these, and nowhere else, lies the real foundation of all authority; build your faith here, and churches and Bibles may come and go, and leave it undisturbed.[18]

At the same time that Higginson was promoting mysticism as the pure source of religious truth, a related movement caught on that emphasized an inner route to healing and health. It was a reaction to the rigorously external focus of modern science-led medicine, and was called the 'Mind Cure' movement or

'New Thought.' It promoted spiritual exercises that brought about inner calm, positive thinking and positive emotions as the keys to good health. Its core metaphysical foundation, as with mysticism, was the belief that the human self or soul is divine in origin, and to live in accordance with it is a force for healing. One of the principal methods that New Thought popularized was meditation, and since that time it has grown exponentially in popularity.[19]

Meditation: Inner technologies of consciousness

Meditation is a broad term that refers to a whole range of techniques that aim to cultivate a state of consciousness that is steadier, calmer and clearer than normal. These techniques can be used to enhance health and well-being, and also to gain insights into the nature of consciousness. The first step in a meditative process is to stabilize attention.[20] Attention generally flits around rapidly between different objects and ideas, switching between past issues and future worries. In the meditative state, it settles down into the present moment.

The Buddhist approach to meditation starts off with focusing attention on the breath. The meditator observes when attention deviates, then places attention back on the breath in a non-judgmental way. There is no trying involved with this kind of awareness meditation – one does not *try* to clear the mind. Any active attempts to try to do so tend to make the mind even busier. One simply sits and pays attention to what is going on inwardly, and over time the mind simmers down and finds itself more in the present as a side effect of being still and non-judgmental.

Other forms of stabilizing meditation, such as Transcendental Meditation and Christian Centering Prayer, place attention upon a mantra that is either spoken aloud or recited inwardly. Moving meditations such as Tai Chi place attention on the moving body, and hence the present moment. The idea in all of these practices is that consciousness becomes still and quiet. The waking mind is like a jar of muddy water that is cloudy due to being constantly shaken, and the first job of meditation is to let the mud settle to the bottom so the water of awareness may become clear.[21] Awareness that is clarified and still becomes the gateway into deeper states of consciousness.[22]

The outer distractions and worries that beset normal life make the stabilizing process of meditation challenging, so periods of silence and retreat can be used to aid the process. Indeed, simply having time away from phones and e-mails can help too in turning the mind within towards stillness, and hence towards that which truly matters to us.[23]

After becoming adept at cultivating an inner state of spacious calm, a person may undertake deeper meditations that have a more focused agenda, such as exploring the sense of self, the nature of reality, or cultivating positive virtues. These include loving-kindness meditation, in which the aim is to focus on sending loving feeling and sentiment to friends and enemies, and meditation on the nature of the self, which involves asking a series of questions inwardly about the self, seeking its true location and nature.[24]

Central to the meditative path is to realize that what we normally think of as 'me' is actually not. This imposter self, or *ego*, is a set of thoughts that one has about oneself, including evaluations of the self as positive or negative, worthy or unworthy, success or failure, or liked or disliked. The waking mind identifies this false self as somehow real, leading to confusions and worries. While meditative traditions agree that this false self is a source of problems and delusion, what the true self is, or whether there is one, is not agreed. That is in large part because the insights gained in deep meditation are ineffable (in other words, they can't be fully expressed in language – more on that in Chapter 8).

During meditation, as preoccupation with the false self becomes less, deeper layers of self and identity become accessible.

The Advaita Vedanta tradition of Hinduism, popularized in the West through the practice of Transcendental Meditation, says that the ultimate realization in meditation is that Atman (the true self) *is* Brahman (God). One of the most famous exponents of this was Shankara, an 8th-century mystic and philosopher, who described how Brahman dwells within all beings as soul, pure consciousness and the ground of all phenomena. To realize one's identity with this is to realize one's identity with the All.[25] Peter Russell, in his book *From Science to God*, provides an account of how during his retreats using Transcendental Meditation, he found that his consciousness did indeed become a pure awareness without any contents at all, and that this provided a sacred insight:

> During these long meditations, my habitual mental chatter began to fade away. Thoughts about what was going on outside, what time it was, how the meditation was progressing, or what I wanted to say or do later, occupied less and less of my attention. Random memories of the past no longer flitted through my mind. My feelings settled down, and my breath grew so gentle as to virtually disappear. Mental activity became fainter and fainter, until finally my thinking mind fell completely silent... When the mind is silent, when all thoughts, feelings, perceptions and memories with which we habitually identify have fallen away, then what remains is the essence of self, the pure subject without an object. What we then find is not a sense of "I am this" or "I am that", but just "I am". In this state, you know the essence of self, and you know that essence to be pure consciousness. You know this to be your true identity. You are not a being who is conscious. You are consciousness.[26]

The Christian tradition frames the mystical realization in ways that show parallels to the Hindu tradition. The philosopher of religion Huston Smith equates the witnessing 'I' of consciousness with the Christian notion of soul.[27] He states that because the 'I' cannot itself be observed, it remains beyond the grasp of science, like a source of light that shines on all things but cannot illuminate itself. Similar realizations of an intangible Higher Self or Divine Self have been depicted variously by mystics who profess no specific religious membership, such as Paul Brunton, Roberto Assagioli and Eckhart Tolle. True to the inner path, these writers all enjoin the spiritual seeker to direct attention inwardly:

Paul Brunton: "It is possible for everyone to find his way back to God because God is present in each of us. But we must begin to search and look, and the right place is within, not outwards. You must first look inwards and find the sacred atom in the heart – the spiritual self within..."[28]

Roberto Assagioli: "The inner experience of the spiritual Self, and its intimate association with the personal self, gives a sense of internal expansion, of universality, and the conviction of participating in some way in the divine nature."[29]

Eckhart Tolle: "When your consciousness is directed outward, mind and world arise. When it is directed inward, it realizes its own Source and returns home to the Unmanifested."[30]

The many dichotomies that define waking life, such as 'everything and nothing', 'self and other', 'inner and outer', are found in deep meditation to be products of thought rather than absolute features of reality. Beyond all of them is a reality that is undivided. So the inner journey paradoxically ends up showing the seeker that the very idea of 'inner' is limited and relative. One of the most well-known contemporary proponents of meditation is Sam Harris, author of *Waking Up: Searching for Spirituality without Religion*. Harris argues that insights gleaned from meditation do not belong to any particular religion. Rather, they are basic and universal truths about the *oneness* of consciousness that run deeper than any singular tradition:

> The deeper purpose of meditation is to recognize that which is common to all states of experience, both pleasant and unpleasant... When you are able to rest naturally, merely witnessing the totality of experience, and thoughts themselves are left to arise and vanish as they will, *you can recognize that consciousness is intrinsically undivided.*[31]

To have this realization of oneness is to move towards a greater sense of connection and love for others, and a deepening desire to relieve suffering of all kinds. So while inner work is usually solitary, its effects are socially connecting.[32,33]

Exercise for developing inner awareness: Meditation on the observer or witness

Meditation is simple, but in its simplicity it can be difficult. The aim of it is to stop aiming. Sit down comfortably on a chair with your feet flat on the ground, shut your eyes and observe what comes into your mind. Observe the thoughts that come up, images that appear, feelings and urges that come to you, and also observe how your attention moves between the outer world and your inner thoughts or feelings. From time to time, you will lose this sense of being aware in the present, and get lost in thought. When you notice that you have lost it, simply note that you have and return to observing. The key is that you don't *try* and do anything. If you are trying something, you will be tense. Just observe when you can, and return to observing when you notice you have drifted away. To help the process along, I recommend getting a meditation app for your phone that provides a timer and plays soothing white noise. Also, wearing an eye mask for the duration of the meditation really helps to relax the brain and prevent distractions.

After the meditation has brought a capacity to stabilize the mind in the present moment, and you feel calm and spacious, then ask yourself the following questions: "Who is this observer that can witness all the thoughts, images and feelings that appear in consciousness?" "How is it that 'I' can experience these phenomena in a way that makes me feel that I am not them – I am the observer of them?" "Where is the 'I'?" "Can the 'I' ever be the *object* of attention or study if it is always an observing subject?" The answers to these questions are hard to express in words, but by entering into this deepest yet nearest of mysteries, you may find yourself inspired to continue the journey into the nature of the self.

At the science-spirituality interface: The mindfulness movement

In the 1970s, meditation in the West was seen as the preserve of hippies and new age travelers. There were few signs on the horizon that it would soon become a key topic of medical and social science, while becoming practiced by millions as a secular practice for well-being. Today, meditation is done in schools, universities, businesses, care homes, hospitals, sports academies and countless more settings. However, in these locations it is usually not called meditation. It is called mindfulness.

The rise of the mindfulness movement can be traced to a specific person and time. It was 1979 when a 35-year-old biologist and student of Buddhism called John Kabat-Zinn began a 10-week meditation intervention for sufferers of chronic pain. The program involved 45 minutes of breathing awareness meditation every day and a short retreat at the end. Kabat-Zinn was convinced that the health benefits of meditation could be enjoyed irrespective of whether it was taught within a religious tradition or not. Crucially, he avoided the words meditation and spirituality. Instead, he settled on the translation of the Buddhist word *sati*, which means mindfulness. In the process, he created a secular idiom for meditation, which proved to be just what was needed to make it palatable to science.

Kabat-Zinn's initial results were positive. The chronic pain patients in his first study reported not only less pain, but less depression, anxiety and fatigue too.[34] However, there was no sudden rush of interest following this pioneering work. By 1990, just 12 papers had been written on mindfulness as a medical treatment. It was after the year 2000 that the research community latched on to mindfulness, and the rise was exponential. Thousands of articles were subsequently published on its effects and its physiological correlates. The rush has not let up – currently about 1,000 articles on mindfulness are being published every two years or so.

Psychotherapy was one of the key areas of uptake. Mindfulness-based stress reduction, mindfulness-based cognitive therapy, acceptance and commitment therapy, mindfulness-based relapse prevention and mindfulness-based eating awareness training were just some of the techniques that sprang up around the mindfulness revolution. Mindfulness then reached some more surprising corners of society. The US military developed mindfulness-based mind fitness training (M-fit) as a form of inoculation against post-traumatic stress disorder. Businesses and banks too latched on to mindfulness as a way of keeping their employees unstressed and thus productive.

Justifying all these cultural uses of mindfulness has been the evidence-based support of science. However, the research on mindfulness has been mostly based on methods that are far from scientifically watertight. Typically, in a mindfulness experiment, a group will undertake a program of meditation and all participants will complete a self-report measure of well-being, depression or some other questionnaire before and after the program, and any change over time will be compared against a control group. Because participants know if they are in the experimental group or the control group, there is the chance that their responses will show a 'demand effect' – in other words, they may respond to the questionnaires as they think they are expected to.[35]

Studies have also been conducted on the physiological effects of long-term meditation on the brain. Results from this work suggesting that meditation practitioners have differences in the corpus callosum, hippocampi in both hemispheres and frontal cortex, compared with non-meditators.[36] This research is important, yet it is difficult to establish if the brain changes found are the direct result of meditation, or due to other behaviors that regular meditators are more likely to do than non-meditators, such as refraining from drinking alcohol.

Another important limitation of research on mindfulness is that it tends to downplay the negative effects that some people experience when trying to meditate. Experimental studies generally look at whether *most* people in mindfulness programs change in some positive way, and make conclusions accordingly. Yet in the minority, it can lead to traumatic memory flashbacks, feelings of anxiety, or feelings of dissociation. These difficulties have been little researched, but there is now an increasing appreciation that mindfulness may have downsides as well as upsides.[37]

Some Buddhists are not comfortable with the mindfulness revolution. They feel that mindfulness has taken meditation out of its context as a practice for promoting ethical behavior via an understanding of the true nature of self, and reduced it to a self-help product. Some have even suggested that mindfulness may be used in some places as a kind of 'cow psychology' – a way of keeping individuals docile despite an unfair social system, or productive within a toxic corporate ethos.[38]

While it is true that mindfulness has become detached from its spiritual roots, a person who practices mindfulness meditation every day for health reasons can't help open up the inner world in ways that lead to insight and growth. As the mind quietens, it becomes receptive to intuitions and feelings from the unconscious that the busy, thought-filled mind blots out. It is no wonder that some people have a difficult time with mindfulness, for the unconscious is a repository not only of our higher wisdom and values, but also of our darker inclinations and neurotic conflicts.

Encountering the unconscious and therapy as spiritual practice

The inner journey that is induced by meditation is by no means a simple path towards peace and love. As meditation allows the mind to calm and stabilize, various emotions, urges, images, memories and intuitions may appear in consciousness that have hitherto been kept at bay in the unconscious. As well as this material being indicative of one's higher nature, it may also be sexual, melancholic, aggressive or fearful in content, as it is precisely these kinds of dark feelings and memories that tend to be repressed within the unconscious due to their unpalatable or disturbing nature.[39] As the light of awareness grows in meditation, so does the circumference of darkness, and the meditator may at times get the sense of becoming both more aware and more vulnerable. The concept of spiritual emergency has been coined to refer to periods on the spiritual path when the process of inner growth and change becomes chaotic or overwhelming.[40]

Jung's theories have shaped how contemporary spirituality

makes sense of the encounter with the unconscious, through his own writings and others who have developed his ideas further, including James Hillman, Anne Baring and Stanislav Grof. In the Jungian tradition, the unconscious is conceived as a realm of powerful forces, energies and personalities, which are invisible to the waking mind but communicate with it via dreams, visions, fantasies and intuitions. It has layers, extending away from the ego and conscious mind into increasingly transcendental realms. The closest layer to the conscious mind is the *personal* unconscious, which contains one's own memories, life story, and inner conflicts. The layer beyond that is the *collective* unconscious, which is a shared layer of mind that exists between persons. It is populated by archetypal figures and symbols that extend across cultures and times, such as the Wise Old Man, the Great Mother, the Hero, the Tower, the Tree of Life, and many more.

All these deeper layers of inner reality exist upon the same unified ground – the *unus mundus* or Original Mind, which for Jung is synonymous with God or divinity. Indeed, the term unconscious is really something of a misnomer in the Jungian approach, for the unconscious is teeming with intelligence and consciousness. It is just that the normal self is usually *unconscious of it*.[41] To enter into dialogue the unconscious and its language of symbols and myth is, says Jung, the essence of the inner mystical path that has been described countless ways across different religions.[42]

For those pursuing the path of meditation, encounters with images from the personal or collective unconscious can be disconcerting and distressing.[43] Dream-like visions can emerge within the meditative state, as the unconscious seeks a new dialogue with an increasingly sensitive conscious mind. Aspects of the self that have been relegated to the unconscious through fear, guilt, shame or anger are often re-awoken as a person develops their inner awareness. These are collectively referred to by Jung as the *shadow*. The shadow may be experienced during meditation or dreams in personified form, as a demonic being, a wicked presence, an evil person or a monster. This shadow side of the human mind is part of the whole, and to seek a more unified state of being, the seeker must integrate it. This means directly facing one's fears and anger, and heading down into the darkness to find what has been hidden there over the years, due to it being appraised as unacceptable or shameful.[44] To come to terms with the shadow means accepting that the line between good and evil runs through every human heart, including one's own. This is a hard but profoundly important lesson, which takes years of committed inner work to fully integrate into life.

Because the process of meditation brings up much into consciousness that is sensitive, challenging and emotionally powerful, therapy is an important process for many on the inner path, particularly the independent spiritual seeker who lacks a spiritual mentor or regular meditation group.[45] A number of forms of therapy explicitly acknowledge their role as aids on the spiritual path, including psychosynthesis, transpersonal therapy, and Jungian psychoanalysis.

In her book *The Inward Arc: Healing in Psychotherapy and Spirituality*, transpersonal therapist Frances Vaughan describes how psychotherapy is an important aspect of the inner journey to awakening and self-transcendence.[46] It allows a person to discover their capacity for self-healing, and to come to terms with the powers that lie latent within their unconscious, in dialogue with someone else who has walked the path. Entering the process of therapy may be a response to some form of inner distress, but it is certainly not an expression of weakness. On the contrary, it is a courageous step to give one's hidden side a voice and so become more whole, or at least less divided.

A key therapeutic tool for exploring the unconscious within the Jungian analytical tradition is dream work. Robert A. Johnson is an articulate exponent of Jung's approach to dream work.[47] He explains that awareness-raising processes such as meditation lead initially to the growing realization of *inner conflicts* between values, urges, beliefs, loyalties, desires, and even of entire subpersonalities that exist within us at the unconscious level. It is by going consciously into these conflicts, seeking meaning in them, and dialoguing with them, that we find the way to higher stages of development. Dreams are one technique that can be used on this challenging quest to reconcile conflicts. Every dream is, says Johnson, a symbolic portrait of one's inner life and the conflictual challenges that are being worked through. The function of every dream is to communicate something to us that we don't know about ourselves, that we are unaware of. To observe dreams carefully, and to learn how to read the symbols in them, forges a powerful link between consciousness and all that exists beyond it.

The four steps to working with dreams that Johnson describes are: (1) Writing down a dream and then noting every spontaneous association that you have with each image, place or person in the dream; (2) Making links between these associations and aspects of one's own life situation, emotions, and problems; (3) Finding an interpretation that ties together the personal meanings from Step 2 into one central idea or insight; (4) Engaging in a practical act or a personal ritual that will bring the message or insight of the dream into physical reality.

In dream work, what matters most is not the outcome but the process. The acts of writing dreams down, paying attention to them, and seeking to interact with them meaningfully, if done over a period of time, bring an expanded inner awareness irrespective of what interpretations you come up with.

Johnson cautions that dream work often involves encounters with the shadow and other powerful unconscious energies. It is therefore best done with a therapist, but can productively be done on one's own too. For those of you who are motivated to engage in your own dream work, I recommend Johnson's book *Inner Work* as an excellent practical manual.[48]

Dual-aspect monism and the four quadrants

To make sense of how the external objectivity of science and the inner subjectivity of the spiritual path can *both* be accommodated as real and valid, Jung adopted a philosophy called *dual-aspect monism*. He drew on the philosophy of Schopenhauer for this task, and integrated ideas from quantum physics too.[49] Dual-aspect monism says that mind and matter exist relative to each other as two inseparable yet different halves of a whole. Schopenhauer describes this as follows:

> The one half is the object, the forms of which are space and time, and through these, plurality. The other half is the subject, which is not in space and time, for it is whole and undivided, in every percipient being... these halves are inseparable even in thought, for each of the two has meaning and existence only through and for the other, each exists with the other and vanishes with it.[50]

Dual-aspect monism asserts that beyond mind and matter, reality at the most fundamental level is neither mind nor matter, but something else that is beyond our direct grasp. Jung called this primary, unknowable reality the *unus mundus*.[51]

A more recent theory of the science-spirituality relationship picks up on dual-aspect monism, and extends it further. Ken Wilber, in his book *The Marriage of Sense and Soul*, argues like Jung that science focuses its attention on exteriors, while spirituality focuses on the interior dimension.[52] The unitary reality that lies behind mind and matter is referred to by Wilber as *Spirit*. All organized entities in the universe (which he calls *holons*), from atoms to humans to galaxies, have both inner and outer aspects. The mind and the brain are simply two different views of the same thing; one viewed subjectively from within, one objectively from without. These two views are intimately related but must be understood using different methods – empirical science for the outer view and meditative contemplation for the inner view.

Wilber goes further than dual-aspect monism in stating that not only does every holon have inner and outer aspects, each also has individual and plural aspects. For example, each atom is made of plural parts, is singular, and is itself a part of a molecule. This pattern goes all the way up through reality; for example, every cell in the body is made of many parts, is a singular whole, and is one of many parts of the body. Everything in the universe is thus *both* many and one, while having inner and outer aspects.

By crossing the polarities of inner-outer with individual-plural, Wilber has developed a framework of four quadrants, as illustrated in Figure 3.1: Inner-Individual 'I', Outer-Individual 'It', Inner-Plural 'We' and Outer Plural 'Its.'

Figure 3.1 Wilber's four quadrants

	Inner	Outer
Individual	I	IT
Plural	WE	ITS

According to Wilber, everything in the universe has these four complementary aspects, and hence can be studied in four different ways. Science and spirituality are complementary in this regard; science tends to focus on the *It* and *Its* quadrants, and spirituality on the *I* and *We* quadrants. The inner spiritual path covered in this chapter represents the *I* quadrant. In the next chapter, the focus moves on to the *We* quadrant of spirituality, in the context of the impersonal-personal dialectic.

Chapter 4

Impersonal – Personal

In his late twenties, Thomas Berry finally achieved his childhood dream of being ordained as a Catholic priest, following four years of study in a monastery of the Passionate Order. Yet rather than celebrating, he found himself doubting his vocation as a priest. To get some perspective on his situation, he packed his bags and went traveling. He hoped that by learning more about the cultures and religions that were omitted from his monastic education, his own life path might become clearer. His subsequent journeys to China, India and the Philippines in the 1940s opened his eyes to the beauty of traditions beyond his own. In China he found the religions of Taoism and Buddhism place a strong emphasis on living in harmony with nature, and this resonated with his own feeling of connection with the natural world. He wrote of one particular inscription in China that became a favorite phrase:

> Heaven is my father and earth is my mother and even such a small creature as me finds an intimate place in its midst. That which extends throughout the universe, I regard as my body and that which directs the universe, I regard as my nature. All people are my brothers and sisters and all things are my companions.[1]

Over the course of exploring these other cultures, he also became increasingly drawn to science, particularly the disciplines of cosmology and ecology. In the end, he decided not to enter the priesthood, and instead to become an academic. Following completion of a philosophy doctorate, he became a renowned authority on Asian religions, Native American religion, and the relationship of evolutionary theory to religion.

Berry was a successful academic but only became famous after he retired. He published several books that set forth an ecological spirituality based on a sense of personal kinship with nature and the cosmos, which were like mission statements for the growing movement of ecological spirituality. In *The Dream of the Earth* (1988) and *The Sacred Universe* (2009), he talks about how science and spirituality are complementary ways of knowing. Science is in his view an *impersonal* approach to understanding the world, which encounters all things (including animals and humans) as objects, and thus necessarily prioritizes the objective. In contrast, spirituality is centered on the experience of *personal* encounters with other conscious subjects (either earthly or divine), for such encounters provide an opportunity for love and compassion. For Berry, personal and impersonal ways of knowing are complementary, for everything in the universe is both subject and object, therefore amenable to spiritual and scientific understandings.

This distinction that Berry made between impersonal and personal encounters was influenced by the book *I and Thou* by philosopher Martin Buber.[2] In Buber's terminology, an impersonal encounter is between *I* and *It*, while a personal encounter is between *I* and *Thou* (or to use the modern English – a personal encounter is between *me* and *you*). Buber argues that the *I-Thou* experience of relationship and mutuality is the primary experience both in theistic experiences such as prayer, and also in experiences of loving relationship with humans and other animals. Buber describes the spiritual profundity of looking into the eyes of his pet cat, and how in that state of connection the isolated self is transcended in a singular moment of relationship. Buber also describes the contrasting power of the impersonal *I-It* encounter in its capacity to separate self from other, and pin reality down into observable objects and facts. It is precisely these attributes that have made it powerful as a tool for science.

The impersonal encounter and science

Through the *I-It* encounter, the scientist encounters the world as a collection of objects that can be recorded, drawn, modeled, discussed and theorized about in ways that need little mention of the person who is doing the observing. The universe as depicted through the scientific lens is thus seen as a large array of objects, and forces that act on those objects. Each observed object should ideally appear the same and have the same properties, no matter who is observing it, or where they are observing it from. In the words of philosopher Robert Solomon:

> Science is taken as a paradigm of rationality in part because of its rigorous disinterestedness. The scientific method is designed to eliminate personal and cultural differences and perspective biases from its studies. It does not matter who did an experiment. It is designed such that it can (one hopes) be duplicated by anyone. Ethical principles and codes also aim at eliminating prejudice and favoritism (especially favoring oneself) in ethics.[3]

By separating self from object, and so taking a detached observational position, the impersonal approach of science permits descriptions of the world that are the same across time, place and persons. In so doing it provides an account of the world that transcends all personal points of view, or what philosopher Thomas Nagel calls a *view from nowhere*.[4]

Any object observed through the impersonal scientific lens shows itself to be a collection of parts. These parts are also made of other parts, and so on, down to the very smallest subatomic particles. The interior structure of objects is thus more objects within objects, the smallest of which according to current science is the quark. Reality can thus be impersonally described in purely 'objective' terms at all scales.

The process of taking an impersonal approach to gathering

knowledge requires training. It is a hard task, for all researchers are motivated by emotions and urges as well as reason and evidence. Taking an impersonal view of *oneself* is particularly hard, as self-perception is typically clouded in various biases, including the tendency to see oneself in an overly positive light.[5]

Exercise for impersonal awareness: Getting '360 feedback' on your personality

Taking an unbiased and impersonal view of oneself takes a lifetime of practice and development. It is a vital skill for the scientist and indeed for anyone else seeking to develop a clear self-understanding. This exercise will help to become aware of any biases or blind spots in your view of yourself. You may notice some resistance in yourself to try it, as it may lead to realizations you may not be happy about. But I suggest you do it anyway.

Ask three people that you know, for example a parent, a friend and a work colleague, to rate your personality on a scale of 1 (strongly disagree) to 5 (strongly agree) for each of the following ten adjectives (which are adapted from a recognized personality questionnaire called the BFI-10)[6]: *reserved; generally trusting; lazy; relaxed; artistic; outgoing; critical of others; thorough; nervous; imaginative.*

Then also rate yourself on these adjectives using the same scale. Next, meet with each of your raters in turn, and ask them what they have scored you, and ask them why. If they have given different ratings, try and explore why this difference in how you come across to others occurs. At the end of the exercise, you will find that you know more about yourself than you did before, and you may find that you have a clearer conception of yourself and your personality than you did at the beginning.

Science uses various means to ensure that the personal perspectives or biases of the scientist can't influence findings. For example in medicine, randomized double-blind control trials are used to test drugs. One group of participants is administered the drug, and another gets a placebo (a harmless sugar pill). The double-blind design means that *neither the researcher nor the patient* knows who has taken the real drug or the placebo. The rationale for not allowing the researcher to know is that science recognizes that many scientists may unconsciously or consciously try to influence the experiment to bring about the desired effect, for they are human and have a personal agenda. They could subconsciously imply to a patient about whether they were taking the real pill or placebo, give the pill to those patients who had the best chance of recovery, or score recovery ratings higher in the group who has taken the pill. The double-blind design ensures that none of this can occur, thus if a positive finding *is* found, it is far less likely to be due to any subconscious influence from the researcher.

One of the most eloquent exponents of conducting science in a rigorous and impersonal way was Johann Wolfgang von Goethe. Goethe was a mystic, poet and writer, but when it came to science he suggested that all personal idiosyncrasies should be left at the door. He wrote an essay in 1792 called *Experiment as Mediator between Subject and Object*, in which he extolls the virtue of the rigorous experiment. Here is an extract from the essay that emphasizes the scientist's need for self-discipline:

As soon as we observe objectively, i.e. free of desire and repulsion, an object isolated from its context, as well as embedded in it, we arrive by calm attention at the faculty to mentally depict and grasp the object's definition, idea or concept. The more assiduously we persist, the sharper become our powers of observation... When the scientific observer attempts to utilize this very power of shrewd judgment to penetrate nature's secrets, seemingly insuperable difficulties begin to loom. All by himself he enters a world where he must guard each step, beware of haste, stay tethered to his purpose, while also heeding the leading and misleading pointers on the way. More or less solitary, neither helped nor hindered by external controls, the scholar-scientist must watch sternly over his own behavior amidst his energetic efforts. One can see that these are stern and strict demands, and one must hope against the odds to fulfill them. No obstacle must impede the attempts to proceed methodically with the inquiry as far as is humanly possible.[7]

This task of maintaining a commitment to rationality, criticality and impersonal disinterestedness is often referred to as being *rigorous*. In the words of the great chemist Louis Pasteur, it is only "rigorous observation of facts" that distinguishes truth from error.[8] This word as used in science stems etymologically from the Latin *rigor*, which means hardness. In reference to the practice of science, rigor represents the qualities of precision, toughness, and reliability, which are juxtaposed to the 'soft' and pliable influences of emotion, superstition or persuasion. This same etymology is why the natural sciences are colloquially referred to as the hard sciences.

The writing style of science is designed to embody the impersonal encounter. Scientific articles often also use something called the passive voice, which involves describing a piece of research without using the pronoun 'I'. For example instead of saying, "I conducted the experiment in a temperature-controlled laboratory," they would say, "The experiment was conducted in a temperature-controlled laboratory." This impersonal linguistic style intentionally excises the researcher as a person with feelings and a point of view. They become a passive receptacle for knowledge reporting. Not all scientific journals now insist on the passive voice, but it is by far the dominant scientific style to this day.

Science writing also maximizes the use of nouns, as nouns convey discrete and observable entities that can be quantified. This means that verbs often get turned into nouns – a process called *nominalization*.[9] A verb, being an action word, always implies an acting agent, while a noun can be presented more impersonally. For example rather than saying, "two molecules were added to the solution," the scientific style would be, "there was an addition of two molecules to the solution," or in a social science paper, "eating healthily" may be described as "a gastric health behavior."

A key source of impersonal rigor in science is the use of mathematics. In a mathematical model of how something functions, there is no personal experience, or even life or death. In his book *Mathematics and the Physical World*, Morris Kline illustrates the impersonal agenda of mathematics by describing the following mathematical problem: A young woman, who can't swim and weighs 110 pounds, falls into a pool from a diving board of 10 feet. How long does it take her to hit the water? The answer can be transferred into algebra, using the equation $d = 16t^2$, if d is the distance from the water.[10] This impersonal, mechanical answer to the mathematical problem does not include the fact that there may be a dead woman in the pool. Death cannot be built into the mathematical formulation, for there is no life or death in equations.

The principle of being oblivious to the suffering or mortality of others is a problem to the impersonal vantage point more generally. It tends to ignore subjectivity and this means that it can be callous. I return to this point later in the chapter. From the perspective of Thomas Berry, whose ideas started the chapter, *every* entity in the universe has a twofold nature as subject and object, with more complex entities having more complex objective forms as well as deeper and more complex subjectivities. Berry concluded that every living thing could be encountered as both personal and impersonal, and that the former is the very essence of the ethical spiritual life.

Personal encounter and the way of spiritual relationship

To use Martin Buber's terminology, an impersonal encounter is between an *I* and an *It*, while a personal encounter is between an *I* and a *Thou*. To acknowledge another being as a *Thou* is to experience them as a conscious being with their own interior point of view. If the other being, in turn, recognizes you in the same way, then the personal encounter becomes an interpersonal one. For Buber, the *Thou* of an *I-Thou* encounter can be a tree, animal, human or the 'eternal Thou' – God. Whoever the other is, seeing them as a *Thou* activates moral concern, for it means they are seen as a being with a conscious point of view, rather than an object. This means that what one does or says can potentially help them or cause them to suffer, and this must be factored into decisions and actions in relation to the other.

In an impersonal encounter, an interaction is one-directional: you encounter and observe the object, but the object does not observe you. In contrast, the personal encounter is a *mutual* interaction between two beings who perceive each other and relate to each other in a reciprocal process of influence; they are both observer and observed. Subject and object thus merge and get mixed up in a higher unity within the *I-Thou* encounter, like two dancers who together make one dance from the intricate blending of their actions and reactions to the other.

While the ideal of an impersonal encounter is accurate observation and disinterestedness, the ideal of a personal encounter is empathic connection and kindness. Desmond Tutu, in his book *No Future Without Forgiveness*, reflects on this in the context of his experiences in the reconciliation process in South Africa. He concludes that friendliness is one of the highest human virtues, and can only occur when individuals meet in

I-Thou relationship as persons.[11] In cases where enemies have been made, reconciliation involves overriding the tendency to dehumanize the enemy, via open dialogue. Tutu contends that where there is genuine two-way *I-Thou* interaction, there will eventually be forgiveness.

Friendly and loving personal relationships provide a context in which a person may feel a spiritual experience of transcending their sense of self, as the *I* becomes part of a *We*. This sense of self-transcendence within relationship is valued across many religious traditions. For example, the poetry of mystical Sufism uses verses about loving relationship to represent sacred truths of oneness. The following verse by Rumi is a beautiful example:

A moment of happiness,
You and I sitting on the veranda,
Apparently two, but one in soul, you and I.
We feel the flowing water of life here,
You and I, with the garden's beauty
And the birds singing.
The stars will be watching us,
And we will show them
What it is to be a thin crescent moon.
You and I unselfed, will be together,
Indifferent to idle speculation, you and I.
In one form upon this earth,
And in another form in a timeless sweet land.[12]

No matter how much intimacy and connectedness is found in an *I-Thou* relationship, there is always an element of mystery in it too, for the conscious point of view of the other person can never be known directly. Within any personal encounter, one thus exists in a state of partial unknowing, never being certain what it is like to be the other, or even if they are conscious at all.

It is hard at the best of times to know how another person sees the world, and it is even harder with animals. The philosopher Thomas Nagel famously discussed the question, "What is it like to be a bat?" and answered that we can never know, no matter how rigorously we study what the bat is made of and how its brain and nervous system work.[13] A bat's sonar-based perception undoubtedly provides an inner experience, but what that would be like is a mystery. Imagining the inner life of advanced non-mammals is even harder. The octopus is an intelligent creature that shows problem-solving abilities similar to many primates, but only a third of its neurons are contained in its brain – each of its eight legs has its own brain-like structure within it. What the octopus must experience with its nine brains is an even deeper *I-Thou* mystery, and one that logically can't be solved by any approach that looks at the octopus from the outside.

Practice for interpersonal awareness: Random acts of kindness

When the current Dalai Lama was asked to distill the spiritual life down to one word, he answered "kindness". To act with kindness is the natural expression of the *I-Thou* encounter; of recognizing that another being is a conscious subject worthy of your care and empathy, and then doing something about it. It is a simple way of focusing attention beyond the self and facilitating a sense of connection with others and the world around you.

One of the more prosocial products of the Internet has been the rise of the *random acts of kindness movement*. There are now scores of websites devoted to this movement, each of which provides ideas and resources for being kind to others in creative and often anonymous or non-reciprocal ways. Examples of random acts of kindness include: sending a friendship letter to a lonely elderly person (see www.silverline.org.uk); giving an unexpected compliment or a message of gratitude to someone you love or admire; leaving a book you have finished somewhere that someone else can pick up; leaving ten pounds in an envelope marked 'open me – I promise to make you smile' with a friendly note inside wishing the lucky person a good day; checking social media for friends who have a birthday today and sending a heartfelt message of positivity for the year ahead; giving a homeless person a small gift such as a hat or some food; or buying a cup of coffee for the person behind you in the queue at a café.

Have a look online and get inspired. Try to do one random act of kindness every day and after one week reflect on how it positively influences you and your life, as well as the others you give to.

In addition to the personal *I-Thou* encounter being the underlying principle of relationship, love and kindness between humans, it is also the essence of experiencing the divine in all religions and spiritual systems that center upon the idea of a personal God (or Gods).[14] A thinker whose concept of a personal God has had a strong influence beyond his own religion was the Jesuit priest and scientist Pierre Teilhard de Chardin. His 1955 book *The Phenomenon of Man* integrated findings from evolutionary science with his belief in a personal God. He proposed that God and the Universe are in fact the personal and impersonal faces of the same thing – the *All*. Just as we can encounter other humans or animals as impersonal objects or personal beings, by simply switching our point of view, so too can we experience the infinite totality of the *All* as either an object or being. If we approach it impersonally, it will appear as the physical universe, but if we approach it personally, we will encounter God, the Oversoul or

Spirit. The personal form of interaction with the *All* provides the basis of a sense of cosmic or universal love, which according to de Chardin is central to the spiritual life:

> A universal love is not only psychologically possible; it is the only complete and final way in which we are able to love... if the universe ahead of us assumes a face and a heart, and so to speak personifies itself, then in the atmosphere created by this focus the elemental attraction will immediately blossom.[15]

A spiritual practice that centers upon the *I-Thou* encounter with the divine is prayer. De Chardin comprehends prayer not as asking for divine intervention, but as a kind *personal communion* with the universe-as-subject. It is a way of regularly reminding oneself that the *All* has a personal as well as an impersonal face, and that a relationship is possible between the part (human person) and the whole (God). The existential psychotherapist and philosopher Victor Frankl refers to prayer in these terms too. In his book *Man's Search for Ultimate Meaning*, he describes prayer as a practice that almost everyone engages in, whether the partner in such dialogue is recognized as God, conscience or an inner higher self:

> Truly, prayer is a 'person-to-person call.' Indeed, it could be considered the climax of the 'I-Thou' relationship that Martin Buber regarded as the most characteristic quality of human existence, that is, its dialogical quality... God is the partner of our most intimate soliloquies. That is to say, whenever you are talking to yourself in utmost sincerity and ultimate solitude – he whom you are addressing yourself may justifiably be called God... And I am sure that if God really exists he is certainly not going to argue with the irreligious person because they mistake him from their own selves and therefore misname him.[16]

Christianity and the other monotheisms have put the personal encounter with a transcendent God at the heart of spiritual practice, but have de-emphasized the validity and importance of a personal encounter with Nature. It was not always thus – early monasteries were almost always placed in auspicious natural locations and early Christian festivals involved outdoor dances and feasts, but with urbanization and the gradual shift of the population away from agrarian jobs, religion moved indoors, and individuals were discouraged from traditional outdoor sacred practices like circle dances.[17] This unnatural split is proving increasingly unsustainable, particularly given that research shows that encounters with nature are the most common source of contemporary spiritual and religious experience.[18]

Animism: The personal in nature

Animism is the view that all of nature is animated by an indwelling personal consciousness that permeates and connects all things and beings. For the animist, every self-organizing entity in the universe, from electrons to butterflies to supernovae, has both an interior point of view as well as an exterior form. Human consciousness is therefore seen not as an enclosed bubble of awareness in an unconscious universe, but as part of a seamless conscious matrix that extends invisibly through the universe.

For the animist, all living things are sacred because they are alive. Spiritual practice involves deep communion with this vital principle of life in other beings and nature herself. In animist communities and groups, particular animals or objects may be selected as particularly sacred, given their importance within the local ecology. A mystic or shaman acts as an interface between the human world and the ecology of nonhuman beings that interact with their human community.[19] He or she is able to translate the nonverbal languages of animals, trees and spirits into a form that others can understand and use to help or heal.

Animism by no means precludes a God – it just places God or

a Universal Spirit at the apex of a conscious cosmos. For example, Thomas Berry, whose work introduced the chapter, was both animist and theist and saw no contradiction in being both. Although animism and theism have traditionally been pitted against another as pagan and religious respectively, there is no necessary separation between the two. There has, however, been a tendency in God-focused religions to downplay the sanctity of nature, and with that, to diminish the value of the feminine.

It seems to be a universal fact that cultures across history have referred to the Earth and to nature as female.[20] Correspondingly, a reverence for the female as symbolized as Mother Earth, Pacha Mamma, the Goddess or Gaia, is integral to animism. Statues of female deity figures that symbolize the Goddess have been found by archaeologists that date back as far as 20,000 BC, suggesting that Goddess religion lasted for around 17,000 years before the rise of the male God religions 3,000 years ago. With the rise of the male monotheisms, the female-led animistic traditions were collectively labeled as pagan and idolatrous. The sustaining personal encounter with Mother Nature and with other living species became considered a dangerous thing. Meanwhile, nature's role to inspire was replaced with religious texts, and the metaphor of father took over, representing not the living matrix, but higher authority and control. The paternal imagery became almost parodic; in Christianity, for example, both God and priests are called Father, and the Pope is called Papa in Italian, which means father.[21]

At the core of much contemporary spirituality are practices that endeavor to recover the ancient feminine connection with nonhuman species, the Earth and cosmos, within the context of reasoned thought and modern values.[22] The ideas that comprise this *new animism* can be traced back to Romanticism, and its love of nature. Romantic poetry of the eighteenth century was replete with animistic imagery and idea, particularly the poems of Wordsworth. For example, in the poem *Lines Written in Early*

Spring (1798) he wrote:

> Through primrose-tufts, in that sweet bower,
> The periwinkle trailed its wreathes;
> And tis my faith that every flower
> Enjoys the air it breathes.

In the poem *Tintern Abbey* (1789) he writes of an indwelling spirit in Nature:

> And I have felt
> A presence that disturbs me with the joy
> Of elevated thoughts; a sense sublime
> Of something far more deeply interfused,
> Whose dwelling is the light of setting suns,
> And the round ocean and the living air,
> And the blue sky, and the mind of man:
> A motion and a spirit, that impels
> All thinking things, all objects of all thought,
> And rolls through all things.

The philosopher Ralph Waldo Emerson, a contemporary of Wordsworth, also wrote at length of experiencing a deep personal communion with nature, famously lauding its power to create a sense of spiritual wonder. Emerson was critical of Christianity's rejection of natural environments for spiritual experience and ritual. He instructed spiritual seekers to spend time in nature, rather than in church, to experience the divine. In his famous essay *Nature*, he describes his own experience of walking through the woods and feeling a sense of personal relationship with the trees around him:

> In the woods, we return to reason and faith. There I feel that nothing can befall me in life, – no disgrace, no calamity (leaving

me my eyes), which nature cannot repair. Standing on the bare ground, – my head bathed by the blithe air and uplifted into infinite space, – all mean egotism vanishes. I become a transparent eyeball; I am nothing; I see all; the currents of the Universal Being circulate through me; I am part or parcel of God... The greatest delight which the fields and woods minister is the suggestion of an occult relation between man and the vegetable. I am not alone and unacknowledged. They nod to me, and I to them.[23]

Henry David Thoreau and Walt Whitman continued this new animism in their writings and poetry in the mid-1800s, as did George Bernard Shaw in the late 1800s. In the 1960s and 1970s, the new animism rose up again as a number of emerging movements, including the hippies, bonded spirituality and environmentalism together within an earth-centered animist belief system. Their spirituality was not about adhering to rules set by a male creator God, but rather about finding a higher harmony with Mother Nature, by cultivating love and compassion that extends beyond the human species to the many other beings with whom we share the biosphere, and to the Earth herself.[24] Through these movements and others, the female has been returning to spiritual life, seeking to rebalance the masculinized monotheisms and their paternal metaphors. Thomas Berry has suggested that the return of the Great Mother will be the defining symbol of our dawning ecological age and the spirituality that underpins it.

At the science-spirituality interface: Panpsychism and Integrated Information Theory

Spiritual animists are by no means alone in assuming that consciousness is distributed all the way through nature. Many philosophers and scientists have come to a similar conclusion, and they refer to this as *panpsychism*. This is the doctrine that consciousness is a fundamental feature of the universe and has always been there. Panpsychism has a rich heritage in philosophy, including Giordano Bruno and Leibniz in the 17th century, Arthur Schopenhauer in the 19th century, and Alfred North Whitehead in the early 20th century. *Emergentism* is the contrary view – it argues that consciousness is the *product* of biology, brains or complex systems, and is not inherent to the stuff of the cosmos. While emergentism is the dominant approach to consciousness in contemporary science, panpsychism is making a comeback, with some eminent scientists proposing that it may be a testable theory.

One of the core arguments given by panpsychists is that it is impossible and illogical to designate a particular moment during the process of evolution when consciousness and subjective experience appeared. If consciousness did indeed emerge at some point, then logically *one* animal must have been the first to wake with an inner life, while all animals that existed prior to this moment had been unconscious zombies. As this first conscious being then reproduced and created all later conscious beings, there would be a very clear dividing line in the tree of life between conscious and unconscious life. But where would this line be drawn? It is not as easy as simply stating that consciousness is present only in those creatures with brains or nervous systems, for there is no clear dividing line between creatures with and without – some insects have brain-like structures, worms have ganglia-like groups of neurons and jellyfish have nerves distributed through their body but no central nervous system.

The panpsychist view is that there is no dividing line; consciousness is inherent to the cosmos, and has developed continuously in ever-changing grades of complexity and internal structure, from the most basic of internal viewpoints to the vicissitudes of human awareness. Brains and nervous systems do not produce consciousness – they amplify it and add complexity and richness to it.

In recent decades, a theory called *Integrated Information Theory* (IIT) has developed a high profile in the science of consciousness, and its originators, neuroscientists Guilio Tononi and Christof Koch, argue that it is a form of panpsychism. The theory states that any conscious experience (such as the one you are having now) is composed of elements that are intricately integrated into one unified experience. This is quite a trick – the information we receive from the world comes through very different channels, so to stick it all together into a unified subjective field requires a very specific set of brain capacities, particularly 're-entrant architecture' (neural pathways that go in both directions simultaneously and so can *interact* in recursive loops).

For IIT the unified nature of the subjective experience is not produced by the brain, but is amplified by it. Consciousness is the inner quality of *all* integrated systems in the universe. This is why IIT is considered by Koch as a form of panpsychism. He writes:

> The entire cosmos is suffused with sentience. We are surrounded and immersed in consciousness; it is in the air we breathe, the soil we tread on, the bacteria that colonize our intestines, and the brain that enables us to think.[25]

The core claim of IIT is that, because integrated information is an absolute requirement for consciousness, you can mathematically compute if something is conscious or not, by calculating a property called Φ, or *phi*, which is a numerical measure of a system's integrated information. From this, predictions can be made about which brain circuits are involved in consciousness (i.e. those that show high *phi*), or whether brain damaged patients who can't communicate are in fact conscious.

IIT does not claim to answer the conundrum of where consciousness comes from, or why a physical system should have the interior experience that it does instead of another. For example, it can't explain why light with a wavelength of 495–570 nanometers appears in consciousness as green, nor whether I can know if your experience of green is like mine. It makes predictions about the physical and computational correlates of consciousness, and accepts that this will never provide an account of the actual experience itself, which exists in a parallel inner kingdom, untouchable by science and its external poking. Koch states his view on this matter as follows:

> ... the mental is too radically different for it to arise gradually from the physical... The phenomenal hails from a kingdom other than the physical and is subject to different laws. I see no way for the divide between unconscious and conscious states to be bridged by bigger brains or more complex neurons.[26]

An interesting implication of Integrated Information Theory is that there is nothing inherent to consciousness that would preclude a computer or machine becoming so, if it was sufficiently high in *phi*. Koch suggests that in view of the fact that the Internet has more transistors than the human brain has synapses, it may already be conscious![27] We will certainly soon be living in close proximity to robots that look humanoid and can learn like humans, and will be faced with the ethical question of whether their intelligence and programmed emotions amount to an inner life. This will be a high-tech extension of an ethical conundrum that sits at the center of the personal-impersonal dialectic.

The impersonal-personal distinction and the 'ethical boundary'

Whether or not another sentient being is encountered as a personal 'you' or as an impersonal 'it' has moral implications. If you encounter a being as a 'you', you acknowledge their conscious subjectivity, hence that they can feel pain and suffering or happiness and pleasure. A personal encounter draws in issues of morality and ethics in how we behave towards the other, for our actions can cause them harm or alleviate it. Objects that are apparently inert and non-conscious will not make one feel moral concern to the same degree; hitting a rock with a hammer is not as morally problematic as hitting an animal with a hammer.

According to the animist philosopher Emma Restall Orr, everyone must draw an *ethical boundary* between conscious subjects worthy of moral concern and unconscious objects on which we can act without any such concern.[28] This view is by no means the preserve of animists. It is fundamental to almost all ethical codes. Sam Harris, who is more atheist than animist, recently argued this very same fact in his recent book on spirituality *Waking Up*. He writes that "we have ethical responsibilities to other creatures precisely to the degree that our actions can affect their conscious experience for better or worse."[29]

For the animist, this link between a personal encounter with nature and an ethical orientation towards her is central to their activism and concern. For example, Starhawk (Miriam Simos) in her book *The Spiral Dance* writes: "If we call the ocean 'our Mother, the womb of life,' we may take more care not to pump Her full of poisons than if we see the ocean merely as 'a mass of H_2O'."[30] Thomas Berry found that expressing gratitude towards nature, akin to a kind of animist prayer, was helpful in developing a personal and prosocial attitude towards the natural world. He was particularly taken with Native American invocations such as this:

We return thanks – first to our mother, the earth, which sustains us, then on to the rivers and streams, to the herbs, to the corn and beans and squashes, to bushes and trees, to the wind, to the moon and stars, to the sun, and finally to the

Great Spirit who directs all things."[31]

For an animist, living in harmony with animals is central to spirituality. Animals are seen as conscious, and consciousness is divine, so to help animals live healthily and avoid causing them undue suffering *is* spiritual practice. Hence eco-spiritual movements have been at the very center of protests against intensive factory farming, animal experimentation, animal product testing and vivisection. Many have adopted vegetarianism as an expression of their sense of kinship with animals and the concerns over mass farming, echoing the vegetarian poet George Bernard Shaw (also a passionate animist) who was quoted as saying that he is friends with animals, and is not in the habit of eating his friends.[32]

For most people though, animals occupy an ambiguous place relative to the ethical boundary. Some such as pets fall on one side of it and so are treated with kindness and affection, and others such as animals bred for experimentation, factory-farmed meat or product testing are sadly on the other side, and are treated in intentionally impersonal ways. The same ambiguity is present in science. Some scientists consider their study of animals to be part of their spiritual quest and their love of nature.[33] Yet the suffering of animals in the name of science occurs on a massive scale.

Currently, around three million animals a year are experimented on in the UK alone, including rats, cats, dogs, rabbits, monkeys and birds.[34] Many of these undergo procedures that bring about intense pain and fear. Philosopher and animal activist Peter Singer lists the categories of harm that are formally listed as committed on animals during such experiments, including 'blinding', 'burning', 'compression', 'crowding', 'crushing', 'freezing', 'heating', 'hind-leg beating', 'isolation', 'punishment', 'radiation', 'starvation', 'stress' and 'thirst.'[35] Laboratory animals are objectified and depersonalized to extinguish personal concern for their individuality and suffering. Singer provides an example of how this objectification is achieved through language, in an advert for guinea pigs in the journal *Lab Animal*:

> When it comes to guinea pigs, now you have a choice. You can opt for our standard model that comes complete with hair. Or try our new 1988 stripped down, hairless model for speed and efficiency.[36]

Many harmful animal experiments are justified by recourse to the moral argument that the suffering of the animals will be outweighed by the benefits that accrue to humans from such research. While this may be a valid argument for some very impactful medical scientific research such as work done with vaccines, unfortunately many animals suffer in the course of experiments that have dubious benefit to humans. Hundreds, if not thousands, of experiments conducted by psychologists on animals using electric shocks and other punishments, have a tenuous positive legacy to science. Most were gratuitous exercises in animal suffering about topics that could have been studied in other ways. Martin Seligman, now known for his work on the psychology of happiness, conducted research in his early career that involved giving powerful electric shocks and other punishments to dogs, to see if they would try to escape or develop a passive response. One of his articles starts: "Dogs given inescapable shock in a Pavlovian harness later seem to 'give up' and passively accept traumatic shock in shuttlebox escape/ avoidance training."[37] Rather than go into detail here about these and other macabre procedures used in animal experiments, I point you towards Peter Singer's book *Animal Liberation*, which is a critical take on science's legacy of animal cruelty.[38]

Scientific researchers are often caught in the paradoxical situation of having to interact with animals impersonally as experimental objects, while noting that they are sentient beings with feelings. A famous story from 1977 exemplifies this, in which two researchers employed at a laboratory tried to liberate two dolphins, Puka and Kea, who were held there. They observed the prison-like conditions that the animals were kept in, including total isolation from other humans and dolphins, no room in their small tanks for swimming, a punishing regimen of tasks to undertake, and limited food rations. Despite being trained to be oblivious to the personal feelings of these dolphins in the laboratory, the researchers could see evidence of self-harm and depression in the dolphins. In the end, their human desire to see these sentient beings freed from suffering won over. One of the two researchers, Mr. Sipman, said in retrospect:

> After much research and soul-searching, I could not expect science to prove what I came to know in my own heart: that these dolphins are as human as you and I. As such, they were no man's property.[39]

The two liberators were found out and tried. The courts were not sympathetic to their compassion. In the eyes of the law, the dolphins were objects, and thus the case was viewed as one of theft. The defendants were told to repay the laboratory the value of the dolphins, as well as being given a probation sentence and community service. Despite the guilty verdict, the case has been taken since as a paragon of the problems and abuses that abound in our relationship with animals. It highlights that if we perceive sentient beings as just objects, then our moral responsibility is abrogated and we become able to commit cruelties and even atrocities.[40]

Not only can animals be depersonalized and objectified, but other humans can be too. People can and do alter their ethical boundary to exclude some other human beings from their moral concern. They deny their subjectivity and individuality and instead treat them as objects to be used, possessed or abused.

As described in *The Objectification Spectrum* by psychologist John Rector, objectification of other people is a matter of degree; it starts with casual or passive indifference to the suffering of others, increasing to an emotional hardening that prevents any empathic connection with others, and culminating in a full denial of conscious humanity to certain groups of individuals.[41] An influential work on the process of dehumanization is *The Lucifer Effect* by Philip Zimbardo. The author makes reference to the personal-impersonal distinction of Martin Buber for understanding the process of dehumanization. He writes:

> In contrast to human relationships, which are subjective, personal, and emotional, dehumanized relationships are objectifying, analytical, and empty of emotional or empathic content... To use Martin Buber's terms, humanized relationships are *I-Thou*, while dehumanized relationships are *I-It*.[42]

Zimbardo states that dehumanized relationships are adaptive for surgeons who must view their patients as objects to cut and mend, and for doctors or therapists who would be overwhelmed with 'compassion fatigue' if they empathized fully with the suffering of their long caseload of clients. Nonetheless, his research shows that dehumanization leads to immoral behavior towards others if it becomes chronic and pervasive. If a person is viewed as just an object or machine, which is the essence of dehumanization, he or she is seen as incapable of feeling pain, so moral disengagement and abuse is justifiable.

Dehumanizing attitudes and behaviors have been at the root of racism, prejudice, violence, sexual abuse, mass violence and genocide. The use of objectivizing language is key to this dehumanizing process. The enemy other is referred to using depersonalizing terms that help to remove moral obligation, such as 'huns', 'japs', 'gooks', 'rag-heads' and 'hadjis'.[43] A

Japanese general reported that his soldiers had had no problem massacring Chinese civilians during Japan's pre-World War II invasion of China "because we thought of them as *things*, not people like us."[44] For this general and many others in wartime, the victims of their killing are objectivized and therefore seen as beyond the ethical boundary.

In summary, science's impersonal mode of encounter is its strength in seeking reliable knowledge, but morally it can be a weakness, for it entails viewing living things as objects rather than subjects, and we don't tend to care for the well-being of objects. The personal approach of spirituality counters this tendency with a commitment to the importance of honoring subjectivity in self and others. It asks us to recognize our moral responsibility to all other conscious beings, and to extend feelings and acts of care and kindness out as far as is possible.

Chapter 5

Thinking – Feeling

In 1850 a passenger ship called *The Elizabeth* set sail from Liverno in Italy to New York. Aboard was Margaret Fuller, who at the age of 39 was one of the most famous women in America. She was a radical intellectual whose assertive campaigning for women's rights had caught the public's imagination. Sadly, after her long stay in Europe, she would never make it back to her homeland. After two months at sea and within sight of the American shoreline, *The Elizabeth* was caught in a hurricane and sank. Fuller, her husband and her child all went down with the ship. As with so many famous figures that die in their prime, her death brought her even more fame. In 1852 a biography of her was released, and for that year it was the number one bestselling book in America. It remained the top-selling biography for a further four years.[1]

Fuller's rise to fame had started when she had met Ralph Waldo Emerson. Emerson was the founder of the Transcendentalist movement, which combined a romantic and idealist form of philosophy with an independent, passionate and mystical spirituality. Emerson was not afraid to speak his mind, and he frequently denounced conventional religion for being chauvinist, dull and stuffy. In Margaret Fuller, he had recognized a natural ally. Her sparkling intellect and self-assurance were a natural foil for Emerson's own iconoclasm. So when the Transcendentalists created *The Dial* journal to spread their ideas, Emerson appointed Fuller as the editor. She had a huge range of knowledge, while also being passionately but independently spiritual.[2]

Amongst Fuller's many contributions to *The Dial*, she wrote about two fundamental forces that in her view underpin all of creation and human life, which she referred to as *Love and Intellect*.[3] The first of these – Love – is a principle of feeling, connection, acceptance and aesthetic experience, and the second – Intellect – is a rational principle of power, exertion, criticism, and knowledge. The former defines spirituality, the latter science. Fuller wrote that all people contain a mix of these principles, and maturity is found in their integration. Similarly, a mature society finds balance through ensuring complementary outlets for Love and Intellect.

In one essay Fuller describes a dialogue between two characters who exemplify Love and Intellect, which she calls *The Poet* and *The Critic*.[4] Throughout the dialogue, The Critic emphasizes the importance of careful and rigorous thinking. His goal is to compare, classify and analyze phenomena using cognition. He says, "I must examine, compare, sift, and winnow; what can bear this ordeal remains to me as pure gold." The ultimate aim of life, from The Critic's point of view, is the advance of reason. In contrast, The Poet relishes the spontaneous flow of feeling, rather than the intellectual analysis that renders art and music devoid of beauty and sentiment. The goal of life from The Poet's spiritual point of view is to cultivate deep and enduring feelings of love and joy, saying: "I do not wish to be heard in thought but in love... I would pour forth my melodies to the rejoicing winds. I would scatter my seed to the tender earth. I do not wish to hear in prose the meaning of my melody." Fuller depicts these two characters as existing in a necessary tension, and yet both as inherent to human life.

In the footsteps of Fuller, this chapter explores how science is defined by reasoned and analytical *thinking*, while spirituality finds its foundation in various forms of deep self-transcending *feeling*. Throughout this chapter, I define *feeling* as a private and subjective sense that can be positive or negative. The word *emotion* is a broader term that includes feeling in conjunction with facial expressions, bodily reactions, and social meanings.[5]

The term *thought* refers to conscious rational thinking mediated by language.

The Enlightenment and the call to independent thinking

Starting in the mid-1600s, and lasting a whole century, the Enlightenment aimed to bring logic and reason to all matters of human existence, including philosophy, ethics, art, music, politics, religion and social progress. The philosopher John Locke (1632–1704), a key figure in the Enlightenment, wrote that "Reason must be our last judge and guide in everything."[6] Half a century later, Immanuel Kant (1724–1804) was still adding fuel to the Enlightenment fire, arguing that it represented a collective shift in human development towards maturity. In an essay on the topic, he wrote:

> Enlightenment is man's emergence from his self-incurred immaturity. Immaturity is the inability to use one's own understanding without the guidance of another. This immaturity is self-incurred if its cause is not lack of understanding, but lack of resolution and courage to use it without the guidance of another. The motto of enlightenment is therefore: Sapere Aude! Have courage to use your own understanding![7]

Enlightenment philosophers saw many obstacles in the way of their goal of universal rationality and independent thinking. On the question of religion, they aimed not to do away with it, but to transform it. They wanted to dispel superstitious vestiges of the past, question credulous habits, and overturn dogma. Instead, they aimed to construct religion around the idea of a rational God whose goal for humanity was to use reason to comprehend his handiwork. In the words of Locke:

Reason is natural revelation, through which God, the eternal Father of light and fountain of all knowledge, communicates to mankind that portion of truth which he has laid within the reach of their natural faculties.[8]

The methods for scientific investigation, developed during the time of the Enlightenment, were founded on its ideal of universal reason and independent thinking. Scientists used rational schemes, mathematics and arguments to make sense of evidence gained from the world through the senses. The thinking processes that were developed then, and have remained fundamental to science ever since, are logical problem-solving, analysis, interpretation, and critical thinking.

Problem-solving in science and the searchlight of thought

All science is problem-solving, and all problem-solving starts with thinking. This message was the essence of the work of Karl Popper (1902–1994), one of the most influential philosophers of science of all time. In his book *All Life is Problem Solving*, Popper describes how every scientific research study can be considered as a problem-solving cycle. Problems do not exist in physical nature, only in thought, so the problem-solving process always starts as a form of thinking. A problem is perceived when a person desires to achieve a goal or ideal state, but has to overcome difficulties to get there.

A scientific problem aims to develop new or better knowledge about some aspect of reality, which can be supported by evidence gathered through the senses and the instruments of science. There are many knowledge-based problems that are not scientific, including metaphysical, mathematical, ethical and formal logical ones. These problems all require evidence, but not of the empirical kind that defines science.

In order to solve a scientific problem, a problem-solving

process is put into operation, the objective of which is to move scientific knowledge towards a more complete state. This process requires a *lot* of thinking to be done before any data can be collected. A researcher must firstly think of a clear *aim* for the research to achieve and consider what *predictions* and *hypotheses* will be tested (Popper was of the view that one should always test hypotheses in science – he called this the 'hypothetico-deductive' method). Then the researcher must plan a method of finding information that will help to solve the problem and test the predictions. This plan must include: a strategy for where and when to collect data; a projected sample of participants or specimens; a design specifying how the experiment or observation will be conducted and controlled; what instruments and/or techniques will be used to collect the data, and how the data will be analyzed.

So, before any research study can be conducted, an intensive program of rational and imaginative thought must first be engaged in. This makes science selective and concentrated from the outset. In reference to this, Popper called his approach the *searchlight theory* of science, to metaphorically convey that reasoned thinking acts like a targeted searchlight to bring problems and solutions into view.[9]

The fact that science is a targeted problem-solving process, guided by the searchlight of thought, provides it with structure and strength. Nevertheless, it also provides science with clear limits, for all the thinking that precedes data collection and analysis means that a scientist is always seeing reality *through* the filter of thought and expectation, which inevitably means a selective focus. In Popper's words:

An observation is always preceded by a particular interest, question, or a problem – in short, by something theoretical... This is why observations are always selective, and why they presuppose something like a principle of selection... An observation always presupposes the existence of some system of expectations.[10]

What that means is that the theories and facts of science are, to a large degree, the product of problem-solving thought. They are *solutions*, rather than eternal truths, and are never perfect. There is thus, in Popper's view, no final answer or end point in science. He wrote: "Science, as it appears in this logical sketch, is a phenomenon to be understood as perpetually *growing*; it is essentially *dynamic*, never something *finished*."[11]

Scientific analysis and interpretation

Not only is there a lot of thinking to do in the early stages of research, but once data has been collected, thinking again comes to the fore with the processes of analysis and interpretation. Scientific analysis is a complex cognitive process in which the researcher breaks data up into *parts and patterns* (the word analysis comes from the Greek word *analuein* meaning to break up or unloose). Firstly, data must be collated into a spreadsheet or database, and the analyst must reduce and condense it down into a summary form. There is no way around this process of condensing the data – it is the only way of producing concise reports from datasets that can include millions of data-points. The selected data must then be integrated into patterns, trends, maps, tables and graphs.

Computer programs that conduct statistical tests are often used during the analysis phase of research to demonstrate associations between variables, or differences between groups. Yet, even when one hands over this activity to computers, all statistical outputs must be interpreted by the researcher in the light of theory and hypotheses, so there is no getting around thinking.

Sometimes the process of analysis can provide facts without

any inference as to *why* something occurred. For example, it was found through statistical analysis that smoking predicted lung cancer, before it was known why.[12] To deal with *why* in any scientific field, one must move past analysis and invoke the more inferential process of *interpretation*.

Interpretation in science involves suggesting an explanation for a finding, which is likely to be shaped by theoretical assumptions that underpin a research study. Often there are competing theories that can explain a finding, so a scientist must evaluate the relative merits of alternatives, and select the one that appears to best explain the facts. The official name for this logical process is 'abduction.'[13] So how does a scientist choose one scientific interpretation over another? Criteria for interpretation include the following:[14]

- *Falsifiability* – A theory or proposition must be sufficiently clear and precise that it could be shown to be false.
- *Logical Argument* – Interpretations must follow logically and rationally from premises and evidence.
- *Coherence* – The favored interpretation should fit within a broad scheme of knowledge, rather than being an isolated or floating theory that is at odds with broader theories and findings.
- *Generalizability* – The interpretation should explain or relate observed phenomena to some kind of reliable cause or fundamental principle, and so be generalizable to other relevant situations and findings.
- *Parsimony* – The interpretation should be as simple and concise as possible, in so doing reducing complexity to unifying principles.

All the above rules of interpretation assume a preference for unity in scientific knowledge. The importance of parsimony is based on the assumption that the universe is orderly, coherent and structured by unifying principles, and so interpretations of it should be elegant and unified too. Why the universe should be ordered in this elegant and rational way is a question that does not have a scientific solution – that is a question for philosophers and theologians. For scientists, it is enough that parsimony works as a criterion for selecting interpretations of how the universe works.

Despite its ambition for coherence, physics has found itself at something of an impasse in its quest for unity, for in quantum physics, radically different interpretations exist as to why quantum effects occur. To illustrate this, I turn now to an experiment that has been puzzling quantum physicists for over a century now. There is still no agreement on what it means.

The double-slit experiment: An ongoing challenge for scientific interpretation

In 1801 Thomas Young asked the question of whether light is composed of particles or waves and designed an innovative way of solving the problem. He shone light through a pinhole and then put a piece of card with two vertical slits in front of it. If light behaved like waves, it should show *diffraction* when it passed through the two slits. This would mean it would spread outwards from the far side of the slits, in arc shapes, like ocean waves passing through the gap in a harbor wall. The arcs of light emerging from each slit would then interfere with each other in the middle where they overlap. That is exactly what he found.

With this, Young had found clear evidence that light behaves exactly as a wave should. He even worked out the wavelength of light as 652 nm (1 nm is 1,000,000,000th of a meter). Subsequent research has shown this to be pretty accurate. It seemed that the nature of light as a wave had been cracked, but in 1901, exactly 100 years after Young's breakthrough, Einstein and Planck showed experimentally that light is made up of individual particles called photons. This then begged the question of why it appears as waves too.

Following the discovery of the photon, the double-slit experiment has been redone using a generator of single photons as the light source. Photons were fired at the slits *one at a time*, so any resulting interference pattern should be impossible because interference occurs from simultaneous waves traveling through both slits at the same time and then diffracting into each other. Photons, as discrete particles, can only travel through one slit at a time, so should not interfere with each other.

As each individual photon was fired, the photosensitive screen registered a tiny dot, showing what appeared to be particles of light. However, the accumulating dots did not show the pattern one would expect of particles passing through slits (i.e. two slightly blurry slit-shaped marks on the receiving screen), but instead made *exactly* the same interference pattern as when a constant beam of light shines through both slits simultaneously. Somehow each photon had traveled through both slits simultaneously as a wave, interfered with itself, yet appeared on the receiving screen as a single dot!

Things became stranger still when a device was set up to measure which slit the photon actually goes through. The detector found that each photon does, in fact, pass through one slit only. Furthermore, as long as there is a 'which slit' detector in place, the interference pattern disappears! The photosensitive plate only registers two areas of photons that correspond to the slits, showing particle behavior. It is as though photons know if they are being watched.

Physicists can no longer find comfort in this effect just being a strange artifact of light. The double-slit experiment has now been conducted with electrons, atoms and molecules fired one at a time at the slits. All show the same pattern of interference and wave-particle duality. At the time of writing, the largest molecule so far to evidence a wave interference pattern is fluorofullerene, which has 108 atoms (or 2,424 protons, neutrons and electrons).[15]

This double slit wave-particle effect has been around for many decades, but physicists still do not agree on what it means. Indeed, competing interpretations are premised on completely different conceptions of reality. I will briefly describe two of the most popular – the Copenhagen Interpretation and the Many-Worlds Interpretation.

The most widely accepted interpretation of the double-slit experiment (and indeed for quantum physics generally) is the Copenhagen Interpretation. Stemming from the ideas of Niels Bohr, it theorizes that a quantum particle such as a photon *when not being measured* does not exist in one state or another, but rather it exists in all of its possible states at once. This is called *quantum superposition*. The total of all possible states of a particle is called its *wave function*. We can therefore never precisely say where a quantum object is until we have observed and measured it, for it is not anywhere in particular until we do. When we do observe or measure an object, there is a *wave function collapse*, at which point the particle takes up a particular state and location. Of the double-slit experiment, it says that the photon is a wave function of superimposed possibilities when not being measured, so it spreads out through both slits. When it is measured using the photosensitive plate, the wave function collapses, making it show as a particle again. According to many theorists who adopt this interpretation, it implies a role for consciousness in physics, for the act of conscious observation collapses the wave function and gives objects definite location.

If the Copenhagen Interpretation seems strange to you, it is positively prosaic in comparison with the Many-Worlds Interpretation. In this interpretation, pioneered by Hugh Everett in the 1950s, there are multiple realities, each of which contains a version of what can possibly happen. There are, it says, different versions of you and me out there in parallel universes. In the double-slit experiment, the photons traveling towards the two slits have two possible future selves – one that goes through the right slit, and one that goes through the left slit. Both of these happen in different branches of reality, but we are only in one of these, so only see one. Or so it seems. When separately fired photons create a seemingly impossible interference pattern, the photon is interfering with the other version of itself in a parallel reality. There is no collapsing of wave functions or effects of observation, just lots of realities in a multiverse of endless possibility that sometimes overlap.

Physicists don't agree on which of these interpretations is correct, and there are now many other interpretations competing to make sense of the double-slit experiment, including Quantum Bayesianism, David Bohm's pilot-wave theory and John Archibald Wheeler's it-from-bit theory. A recent survey of physicists in 2013 found the Copenhagen Interpretation is still the most popular, with 42 percent opting for it. 18 percent supported the Many-Worlds Interpretation, and the remainder chose other interpretations.[16] So it is clear that there are radically different ideas within science about what kind of universe we are living in. To decide between these competing theories, or indeed to evaluate the merits of science generally, an important tool is critical thinking, and we turn to that now.

Critical thinking in science

An essential skill of the scientist, and a key aspect of scientific education, is the practice of critical thinking.[17] To think critically is to rationally seek the flaws and limitations in one's own research and theorizing, and to do the same with the work of others too. Critical thinking thus requires a general attitude of skepticism towards ideas that are presented, including one's own. A good scientist, following best practice, should constantly debate with her conclusions and try to prove herself wrong.[18]

Such humility, while laudable, is not always manifest in science. As with all human beings, scientists are prone to take a particular view on the world and seek confirming evidence, while avoiding contradicting information. This is a phenomenon called the confirmation bias.[19] It manifests in science as a tendency to support theories despite accumulating evidence against them, and a tendency in science to assume that existing theories are beyond doubt.[20] However, scientists often do still admit they are wrong when presented with disconfirming evidence, and make resulting changes to their theory or method in light of such outcomes, and this gives science the flexibility and power to keep developing.

Critical thinking acts not only as a guard against a scientist's own irrationality and bias while doing research, but it also acts as gatekeeper in the process of scientific publishing, via the process of peer review. When a study is written up and submitted to a journal, a selected group of academics are chosen by the editor to give their judgment on its quality, using their skills of critical thinking. This process aims to ensure quality control in science, but given the volume of published scientific articles nowadays, it is hard to maintain with total consistency. According to a recent count, 28,000 scientific journals now churn out 2.5 million articles a year, and sometimes the peer review process fails to ensure that articles are credible and rigorous.[21]

Several frauds have recently been intentionally perpetrated to illustrate the flaws in the peer review system. A scientist called Cyril Labbé created a series of fake papers under the fictitious name of 'Ike Antkare.' He devised a computer algorithm that generated scientific-sounding nonsense and used this to create the papers. An example nonsense title was "Developing the Location-Identity Split Using Scalable Modalities." By optimizing the way that academic search engines count citations, Antkare temporarily became the 21st most cited scientist in the world, above Einstein![22]

In another sting, biologist John Bohannon submitted a deliberately flawed paper about an anti-cancer drug that had a deeply flawed method and false analysis to 304 journals that use the peer-review process. He used a fictional name and a non-existent institute at the top of the paper. Despite the fictional author and erroneous content, more than half of the journals accepted the article for publication. Only 36 of the 304 received peer reviews recognizing the paper's scientific problems.[23] These stings show how science is constantly vulnerable to a loss of critical control, and how critical review is a crucial way that rational thinking contributes to science's progress and credibility.

Exercise for critical thinking: Cultivating the capacity for self-criticism

Exercises in self-criticism help to develop a clear awareness of the merits and pitfalls of your own ideas, opinions and beliefs. This exercise is designed to cultivate this capacity. The task is to debate *with yourself* about a contentious topic. Think of an issue about which you feel strongly, and list the arguments for your position on it on the left side of a piece of paper. Then draw a line down the center of the page and set out counter-arguments to each of your points on the right. To help with clarifying arguments and counter-arguments, have a look online at websites that weigh up both sides of the debate.

If you can't think of an appropriate issue to debate with yourself, I suggest you think about the dispute on the ethics of assisted dying for terminally ill patients, and decide if you are for or against. There are solid arguments for both sides of the debate, and at the time of writing some countries allow it and others do not. I have listed some arguments for and against it in the endnote reference given here.[24]

You can replicate the structure of this exercise with

any big moral or philosophical issues. By doing it with various topics, and becoming accustomed to playing devil's advocate with yourself, you will become more humble about your own capacity for knowing truth, while becoming more empathic to those who disagree with you.

Spirituality and the balance of feeling: Eudaimonia and equanimity

In contrast with science's emphasis on reasoned thinking, the aims of spirituality frequently emphasize the cultivation of certain feelings.[25] The essential goal of spirituality, say authors such as George Vaillant and Roger Gottlieb, is to lessen feelings of anger, hatred and despair, and to increase feelings of love, joy and peace in self and others.[26,27] This focus on feelings reflects the importance that spirituality places on well-being and reducing suffering. It also relates to the conviction that feelings and intuitions can convey insights about mysteries that are too elusive to be squeezed into the rules and concepts of thought.[28]

Spirituality as a route to well-being has a rich heritage that goes back to the New Thought movement, which became highly popular in Europe and the USA during the late 1800s. The central premise of New Thought was that spiritual practices bring not only insight and meaning, but also benefits for mental and physical health and well-being. New Thought writers such as Ernest Holmes, Florence Scovel Shinn and Henry Wood argued that to have insight into the true nature of the self brings joy and calm, which in turn brings a healthy body and mind. This link between spiritual practice and well-being has remained a key pillar of spirituality ever since.

New Thought drew on many sources, including Aristotle, Stoicism and Buddhism. Aristotle promoted *eudaimonia* as a kind of spiritually-informed well-being found when emotions are in harmony with reason, relationships and ethics. The word literally means *in good spirits* – a compound of the Greek words 'eu' meaning good, and 'daimon' meaning spirit. Central to Aristotle's view of eudaimonia is that good and bad feelings need to exist in balance for a good and virtuous life. We should not strive to be happy all the time, but should rather aim for a balance between positive and negative. In the *Nichomachean Ethics*, he writes of this as follows:

> Both fear and confidence and appetite and anger and pity, and in general pleasure and pain, may be felt both too much and too little, and in both cases not well; but to feel them at the right times, with reference to the right objects, towards the right people, with the right motive, and in the right way, is what is both intermediate and best, and this is characteristic of virtue.[29]

The path of eudaimonia, properly pursued, involves a growing tendency towards positivity and compassion, while mindfully bearing the pain and hardships of life's journey. Happiness is seen as part of life's tapestry of feelings, and should not be clung to too tightly. Following on from Aristotle, the philosophical school of Stoicism, championed by Marcus Aurelius, promoted equanimity as the key to a life well lived. Equanimity, like eudaimonia, is all about balance; it involves remaining calm and centered despite the ups and downs of life, particularly through the application of reason and the practical wisdom of only trying to control that which can be controlled.

A century before Aristotle and the Stoics were promoting emotional balance as the key to life in Ancient Greece, in the East the Buddha was disseminating a similar message. The Buddhist conception of equanimity involves keeping aware of feelings, and not getting too attached to outcomes, given that nothing is permanent or fixed. Everything comes and goes, so one should take good and bad fortune as tied together in the flux and flow of life. The important thing is to be kind, to keep up a practice of meditation, and to avoid aspirations to be perfect. It is precisely these aspirations that tend to cause suffering in the first place. The current Dalai Lama, in his book *The Art of Happiness*, translates these Buddhist ideas into a contemporary idiom. He writes of the importance of purpose, warmth, kindness, and compassion to the spiritual life, and argues that these human qualities matter more to a meaningful life than beliefs and metaphysics.[30]

The Dalai Lama's writings are anchored in the Buddhist belief that alleviating suffering is the highest human aim. This is a challenging goal indeed, for research shows that people are inclined to feel negative feelings more than positive ones and dwell on negative memories more than positive ones.[31]

While positive feelings tend to be more frequently reported within spiritual experiences than negative ones, painful feelings do have important healing roles to play too in spirituality. Mystics often report going through a *dark night of the soul* on their journey to spiritual growth, as something that is integral to the path rather than to be sidestepped or erased. Not to face one's fears and darker inclinations can lead to a kind of smile-covered denial of the darker side of human nature. Miriam Greenspan, in her book *Healing Through the Dark Emotions*, sets forth why she believes that grief, fear and despair can all be spiritual teachers in their own right. The message contained within them is to change and move on; to leave bad habits or destructive patterns behind and to grow. However, negative feelings only work in this way if they are mindfully attended to for the tacit knowing they contain, rather than suppressed or hidden.[32]

In summary, positive and negative feelings are ultimately interlinked as parts of the same polarity, so one cannot hope to have one without the other. An elegant poetic depiction of this interrelation is conveyed by William Blake in his poem *Auguries of Innocence*:

Joy and woe are woven fine,
A clothing for the soul divine;
Under every grief and pine
Runs a joy with silken twine.
It is right it should be so;
Man was made for joy and woe;
And when this we rightly know,
Safely through the world we go.[33]

Feeling and the body: The importance of movement exercises

Many of us live in ways that are increasingly disembodied. With every passing year, the average amount of time spent in front of digital devices and TV increases. Academic estimates suggest the average adult in many Western countries now spends 8 to 10 hours a day in front of a screen. This growing screen obsession brings with it an increasing tendency to ignore what is going on in our bodies, and that includes feelings. It is surely no accident that as our lives have become more sedentary and static, movement-based spiritual practices have risen rapidly in popularity. They are the perfect antidote to the digital age.

Spiritual practices that center on physical movement can be broadly categorized into two types: (1) structured form-based practices and (2) spontaneous dance-based practices. Your task, if you have not already done so, is to try at least one of both kinds.

Form-based practices such as hatha yoga, tai chi or paneurhythmy involve learning and perfecting a pre-designed set of movements or postures. The physical forms that comprise the techniques are considered to be sacred in themselves, in that they facilitate the healthy movement of physical energy and/or subtle energy (called *prana* or *chi*, depending on the philosophy). This in turn creates a harmony between the mind, body and physical world.

Dance-based practices such as Five Rhythms (5Rhythms), Biodanza, Movement Medicine or Shakti Dance involve dancing in a spontaneous and expressive way to music or drumbeats. These exercises are an open invitation to allow one's feelings a full and vital expression through the body within a safe social space. The ecstatic feeling of deep connection and openness that such activities bring about is not only eudaimonic but also an insight into how deeply connected we truly are when we can move beyond our self-conscious egos.

I encourage you to try at least one form-based and one dance-based movement practice during your spiritual explorations. Both kinds bring powerful but distinct kinds of experience that help to develop the discernment and control of feeling that is central to spiritual development and good health. As with learning a language, doing a movement practice just once will not lead to results. The benefits such practices bring, and the depth of experience that one can reach through them, builds with time. So whichever you try, stick with it initially for at least a couple of months to give yourself a chance to observe meaningful change. You can then decide whether it should become a regular fixture in your life.

Ecstatic experiences: Momentary self-transcendence through bliss

To experience ecstasy is to overcome one's normal sense of separateness and self-consciousness through feelings that are so powerful that they allow a temporary transcendence of ego. Ecstatic feeling is described using words such as euphoria, bliss, ecstasy, delight, exaltation, rhapsody, passion and transportation. These kinds of strong feelings bring a sense of deep connection between self, other and world, and of an insight into the flowing unity that all things express.

Ecstatic feelings can be cultivated through meditation, worship, and sacred forms of trancing that use dancing, drumming, chanting or singing. Trance in this context is not a reduced sense of consciousness, but an expanded version that is less centered on self-image. Judith Becker, in her book *Deep Listeners*, reflects on the role of feeling and emotion in spiritual trance:

> Trancing in religious contexts draws on emotion, depends on emotion, and stimulates emotion through sensual overload: visual, tactile and aural... propelling the trancer... to the feeling of numinous luminosity that encapsulates special knowledge not accessible during normal consciousness... There is a joy in the pure bodily experience of strong arousal, a life-affirming quality of feeling truly alive that both deep listening and trancing can enhance.[34]

In the contemporary spiritual landscape, a popular form of ecstatic practice is Five Rhythms (5Rhythms), devised by Gabrielle Roth in the late 1970s at the Esalen Institute in California. Roth drew on a wide range of influences – the dances of Sufism, the rhythm and chanting of shamanism, and new ideas about emotions from psychology and psychotherapy. In Five Rhythms, participants dance communally to music that conveys five rhythmic styles: *flowing, staccato, chaos, lyrical* and *stillness*. The facilitator selects the music and guides the participants. The practice is unstructured and spontaneous – there are no set moves. No intoxicants are allowed. Dancers are encouraged to move in any way that they feel the music moves them, and periodically are asked to dance with a partner and in groups. Through such practices, the Western tendency to move in stiff, predictable and structured ways is overridden, and the dancer moves into a more fluid state of embodied being, and more directly into the feelings and emotions that reside within the body.

Roth saw Five Rhythms as a way of bringing mind, body and spirit together in ecstatic union, which she described as "a state of total aliveness and unity."[35] In her book *Maps to Ecstasy*, she emphasized the profound importance of feeling to spiritual development, and the role of dance in unlocking old feelings from hidden places. She was of the view that all emotions must flow outwardly through the body to be healthy, yet our culture often prevents such emotional expression, favoring suppression and control instead. She wrote that: "Freeing the body leads inevitably to freeing the heart... In dancing and singing, you are discovering and releasing the energy of the emotion, allowing it to flow through and out of you."[36]

Five Rhythms is now practiced all over the world. In my own city of London, there are more than twenty weekly classes. There are also many other ecstatic dance practices that have become internationally popular, including Biodanza and Movement Medicine. They all serve the need in Western culture to experience transcendental feelings in safe and social spaces, and so to reach into those oceanic states of embodied consciousness that are, for many, a defining feature of the spiritual life.[37]

There is, however, a lingering suspicion of ecstatic feeling in the Western world, which reflects the continued concern that strong passions may overwhelm reason. This apprehension is not entirely misplaced. If the spiritual life becomes overly focused on the cultivation of ecstasy, an attachment to re-experiencing the 'high' feeling can be created, which in turn creates an unstable ebb and flow between ecstatic experiences with no firm foundation and no ostensible social purpose. While ecstasy really *does* bring insight into an underlying unity and connectedness that the mind is normally blind to, it is neither the whole picture nor the whole point of spirituality. Another more subtle, but no less profound, kind of spiritual feeling is the aesthetic sense.

Aesthetic experiences and the feeling of beauty

Aesthetic feeling can be experienced in response to an external sensation such as landscape, architecture, artwork, music, physical attributes, or an internal experience such as a dream. For the poets, composers and artists who were part of the movement of Romanticism, aesthetic feeling was the central pillar of their spirituality and their path to an expanded consciousness. For the Romantics, to behold beauty was the primary expression of being alive and conscious, and art was the essential means for achieving that aliveness. Wordsworth described poetry as "the spontaneous overflow of powerful feelings", while the composer Liszt asserted that "art is for us none other than the mystic ladder from earth to Heaven – from the finite to the infinite – from mankind to God: an everlasting inspiration and impulse towards redemption through love!"[38] The aesthetic reaction was for them an intuitive form of knowing or truth, which transcends conscious thought and language. It is telling of how our culture has verbally linked the idea of aesthetics with consciousness, that its antonym, anaesthetic, is now used medically to refer to that which makes us unconscious.

The English Romantic poet Keats famously wrote that beauty *is* truth: "I am certain of nothing but of the holiness of the Heart's affections and the truth of the Imagination – What the imagination seizes as Beauty must be truth."[39] Aesthetic objects typically have a mathematical order that is not consciously accessible or communicable. For example, faces and geometrical figures that are considered beautiful across cultures often show the golden ratio Phi (1.618:1) in multiple aspects. Artistic compositions that are considered aesthetic have definable mathematical ratios too, as do beautiful musical harmonies. Hence the aesthetic experience speaks of an order and elegance that is intelligently woven into reality itself.

Aesthetic feeling can be positive, but not always. It can also manifest as sadness and melancholy. The Romantic poets and composers dwelt extensively on this, seeing much truth and beauty in certain kinds of sorrow. For example, Coleridge wrote *Dejection: An Ode*, Keats wrote *Ode on Melancholy*, and the composer Chopin wrote *Tristesse* (Sadness), all on this trope. They found that aesthetic melancholy was a deep and wonderful thing. In depression, tears are an expression of hopelessness, but in aesthetic melancholy, tears become a "hallowed stream" in which the heart awakens to life's brevity and fragility.[40]

There is a danger of equating spirituality too closely with aesthetic feeling, and in so doing mistaking the beautiful part for the whole. If beauty becomes a golden standard for the presence of the sacred or the true, this can lead to an avoidance of the ugly and broken in spiritual contexts, and to a commercialized 'glossy' spirituality based on images of attractive healthy people doing things in beautiful places. In contrast, the next kind of spiritual feeling – the sublime – is far harder to commercialize, being a potent mix of pleasure and pain.

The sublime, awe and the numinous

An important work of comparative religion that drew the idea of the sublime into contemporary spirituality was *The Idea of the Holy* by Rudolf Otto.[41] In this book, holiness and sacredness are described as manifesting through feelings so powerful that they silence the intellect with their significance and magnitude. Otto called this powerful mix of fear, thrill, excitement, love and humbling mystery the *numinous*. It is, he says, at the heart of all valid religions, and the essence of spiritual art and music too. It is the essence of what others have called the *sublime*.

To experience the sublime or the numinous is to encounter a powerful and paradoxical feeling in which positivity and negativity mix in an experience of overwhelming power.[42] The feeling that best captures the paradoxical nature of the sublime is *awe*. In English, the two sides of awe are represented in the adjectives *awful* and *awesome*. The awful refers to the side of awe in which we feel fear, terror, suspense, a sense of being overwhelmed, and the potential for peril. The awesome is the converse side of awe, containing feelings of wonder, astonishment, excitement, reverence, and veneration.

This juxtaposition of positive and negative feeling in the sublime is an ancient motif in relation to the divine. In Christianity, God is paradoxically to be both loved and feared. In the Bhagavad Gita, Arjuna is shown the true nature of divinity and says, "Things never before have I seen, and ecstatic is my joy; yet fear-and-trembling perturb my mind." In Hinduism, the sublime is personified in the figure of Kali, who is the Goddess of both love and destruction – both nurturing and terrifying. The sublime has also been a topic of philosophical discussion, and Kant wrote of its contradictory nature as follows:

> The feeling of the sublime is at once a feeling of displeasure, arising from the inadequacy of imagination... and a simultaneous awakened pleasure, arising from this very judgment of inadequacy of sense of being in accord with ideas of reason...[43]

The sublime experience brings a sense of confrontation with a mystery that transcends reason and our capacity to make sense of it, in a way that is both frightening and exciting. The overwhelming sense of sublime feeling differs from the orderliness of aesthetic experience and its harmonies.[44] Artists seeking to convey the feeling of the sublime have typically selected natural vistas that are threatening, looming, vast and obscure. Famously, Turner's images of storms at sea have been considered as classic depictions of the sublime, in that they capture scenes that are mysterious, vast and far beyond human control.[45]

Events that for many people induce a paradoxical sense of the sublime are birth and death. The birth of a child brings a profoundly complex mixture of pain, elation, fear, hope, and love, and research shows the event often induces a sense of the sublime in parents, and is frequently reported to be a spiritual experience.[46,47] At the other end of life's journey, grief following bereavement can show clear features of a sublime encounter. In her book *Healing Through the Dark Emotions*, Miriam Greenspan describes the experience of losing her baby son Aaron, and how she found an opportunity for spiritual growth as her ego's striving for status and self-esteem was crushed by the grief.[48] The experience was both intensely painful but also an opening to a more compassionate and connected way of living in which she found she could embrace the truth of human vulnerability while opening to the awe of mystery.

In the deeply affective experiences of ecstasy, beauty and sublimity, there is a deep-felt connection between self and other, in which a sense of higher unity and love arises. In all such feelings, self-consciousness and its attendant insecurities lift like a fog, allowing what seems like a purer perception of reality.

All such feelings can be construed as aspects of love, if love is understood as the unifying principle that seeks expression through all connecting spiritual feelings. This kind of spiritual love even includes its opposite, as Ram Dass described in his account of being in a mystical state of consciousness:

Such unbounded spacious awareness contains an intense love of God, equanimity, compassion, and wisdom. In it there is openness and harmony with the whole universe. Beings whose awareness is free to enter into the ocean of love that has no beginning or end – love that is clear like a diamond, flowing like the ocean, passionate as the height of the sexual act, and soft like the caress of the wind. This is the all and everything. It is the love that includes hate, for it is beyond polarity. It is the love that loves all beings.[49]

Experiences such as this do not conform easily to thought and language. An open pluralism of interpretation is thus inevitable when discussing spiritual feelings. This plurality of understanding needn't undermine the unity of the experience and its attendant feeling. As the Unitarian minister Ferenc David said, "We need not think alike to love alike."[50] Nevertheless, all spiritual experiences, no matter how far removed from thought, must be brought into some system of understanding if they are to be made sense of and communicated to others. Indeed, clear reflection and critical interpretation of spiritual experiences are the marks of an adult spirituality in which feeling is properly bonded with thinking.[51] Without this, strong feelings can become an opening through which childish or regressive superstition may spill. On the flip side, too much critical thinking tends to extinguish spiritual feeling, so any process of interpretation and reflection must be done lightly, with the acceptance that the numinous and ecstatic territory of spiritual feeling does not conform fully to the hard edges of critical thinking and propositional logic that science requires.

At the science-spirituality interface: Intuition and deep feeling in research

Leaf through an edition of *Science* or *Nature*, or any other scientific journal, and you will find dispassionate accounts of the thoughts, actions and observations of the scientist, but no mention of the intuitions that helped develop the ideas, nor of the feelings and passions that motivated their efforts. Feelings and intuitions are not accepted as part of the scientific method, and may even be considered sources of bias. This is why they are generally omitted from the formal reports of research.

The autobiographies and nontechnical works of scientists provide an outlet for them to discuss the emotional side of their work in a way that formal articles do not. These personal accounts frequently show the emotional thrill of doing science, which lends its own spiritual meaning. For example, nuclear physicist Lise Meitner has described science as a deeply emotional activity – one that brings wonder, awe and joy.[52] The renowned astrophysicist Carl Sagan also wrote that science engenders deep feelings of awe and reverence in him, and that it is these that make it a spiritual experience:

> In its encounter with Nature, science invariably elicits a sense of reverence and awe... Science is not only compatible with spirituality; it is a profound source of spirituality. When we recognize our place in an immensity of light years and in the passage of ages, when we grasp the intricacy, beauty and subtlety of life, then that soaring feeling, that sense of elation and humility combined, is surely spiritual. So are our emotions in the presence of great art or music or literature, or of acts of exemplary selfless courage such as those of Mohandas Gandhi or Martin Luther King Jr. The notion that science and spirituality are somehow mutually exclusive does a disservice to both.[53]

As well as the feelings of awe and wonder, intuition is important in the work of many scientists. It can be conceived as the mediator between the unknown and the known, or between the unconscious and conscious. When a scientist has an intuition, it comes as a *feeling or hunch* about something without knowing the reasons why, and this kind of inarticulate hunch is regularly the first step in a scientific breakthrough.[54] Intuitions are not wholly spontaneous; to paraphrase WI Beveridge in his book *The Art of Scientific Investigation*, intuition is like a dim light that illuminates a darkness high above us, but only when enough knowledge about a topic has been acquired that a person has been raised upwards to sufficient height to peer into that lofty darkness.[55]

Science has no formal training in cultivating intuition, so it is to spiritual exercises that a person can turn to cultivate practices in unknowing and openness. All exercises for developing intuition involve quietening the din of conscious thought. This can be done through exercise, meditation, ecstatic dance, or creative forms of contemplation such as drawing or doodling. Absorption in these kinds of embodied activity, while incubating an intent in the unconscious to solve a problem, enables the first intuitive inklings of an answer to be noticed. Given persistence and time, these will crystallize into a clear solution. Einstein went so far as to suggest that it is *only* intuition that can facilitate major theoretical advances in physics, for it is only intuition that can truly plumb the depths of the unknown.[56] Discovery in science through the application of intuition thus requires humility, diligence and a mind that is open to mystery. As the physiologist Claude Bernard wrote: "Those who do not know the torment of the unknown cannot have the joy of discovery."[57]

Chapter 6

Empirical – Transcendental

At the age of 35, Evelyn Underhill was in a deep personal crisis. Having studied at King's College and considered an academic career, she had followed the expectations of an Edwardian middle-class woman, and had become the country wife of an English barrister. Yet she knew deep down that it was not enough for her and she was destined for something more. Her crisis had stirred a spiritual search, inspired in part by meeting the Indian sage Rabindranath Tagore when he came to the UK on a lecture tour. She started reading avidly on religions East and West, and spent several extended periods in spiritual retreat. This culminated in writing a book entitled *Mysticism: A Study of the Nature and Development of Man's Spiritual Consciousness*, published in 1911. It became a surprise bestseller, and she subsequently became a renowned figure in the field of mysticism.[1]

Underhill's thesis is that mysticism exists as a unifying theme in all religions, and is both their source and purest manifestation.[2] The mystic develops a capacity to expand consciousness and so become aware of a *transcendental faculty* that remains beyond the threshold of awareness for most people. This faculty brings higher knowledge through feeling and intuition, and also an inner perception of spiritual realities.[3]

Underhill distinguishes science and mysticism by associating the former with *empirical* knowledge of the world based on information from the physical senses and the instruments that extend them, while in contrast mysticism seeks a *transcendental* reality that is beyond the reach of the physical information channels of the body, but can be encountered inwardly in certain states of consciousness. The mystical revelation is, in her words, a direct experience of "levels of reality which in their wholeness are inaccessible to the senses: worlds wondrous and immortal, whose existence is not conditioned by the 'given' world which those senses report."[4] Following Underhill's lead, in this chapter I look at the empirical-transcendental dialectic as a basis for distinguishing science and spirituality, including where the two overlap in transpersonal psychology and the study of near-death experiences.

Science, empiricism and the senses

Science is founded on the assumption that the senses are trustworthy sources of information about reality. This was not always assumed. The medieval worldview that science was rejecting had at its core the Great Chain of Being – a hierarchical conception of reality that places the Divine at the top and matter at the bottom of the pile.[5,6] Within this hierarchical framework, the exploration of the material world via the senses was not widely thought to be a worthy enterprise because it was using the lowest and least perfect level of the human being (the body and its senses) to look at the lowest and least perfect order of reality (the material world). The way to truth was through revelation, reason and mathematics, none of which require empirical evidence.

The early scientists took a new approach to the material world, believing it to be a canvas for God's handiwork, within which signs of his rational order and infinite intelligence could be found. Of all the bodily senses, it was *vision* that dominated science in those early days, and it has ever since. Observation with eyes, either directly or through telescopes and microscopes, was from the outset the primary means by which the outer world was encountered and measured by scientists. Hence we find the metaphor of light widely used to refer to science's method and discoveries. Scientific discoveries were part of the En*light*enment and were described as illuminations. The poet Alexander Pope, a contemporary of Newton, adopted this metaphor when he wrote: "Nature and Nature's laws lay hid in night: God said, 'Let Newton be!' and all was light." Also linked to vision is the metaphor of science as a *window* on to reality. For example, physicist Cecil Frank Powell wrote: "Any device in science is a window on to nature, and each new window contributes to the breadth of our view."[7]

At the same time that science rose to prominence, the philosophical movement of *empiricism* also gave the senses a key role in knowledge. Developed initially by John Locke (1632–1604), it proposed that all knowledge is gained by individuals during their lifetime by way of sensory experiences. From the empiricist point of view, a man does not become learned by spending his life in a monastery or an ivory tower, but rather through rich and varied encounters with the world via the senses. We are like a blank slate (or in Locke's words a *tabula rasa*), and the senses are the paintbrushes. The richer and broader the sights, sounds, tastes and smells that can be gathered, the more developed the mind becomes, for every new sensory impression brings new learnings.

One of the key tenets of Locke's empiricism is that the senses are clearly limited in their scope to provide a full picture of reality, and we are ignorant of what and how much lies beyond them. Hence he argued that we must be agnostic about transcendental realities and accept that we are locked in a sensory bubble that defines the limit of our reality. Scientists took a more practical approach to the problem of the limits of the senses. Rather than accepting human limits, they decided to try and enhance the senses using technology, to increase the range of information that can reach us from the world, and in so doing to draw what was previously thought to be transcendental into the empirical domain.

Science and improving the senses: Magnification, measurement and mapping

Science has always been a synergy of man and machine. Through the devices that have extended the reach of the senses, scientists have become able to sense things that our raw primate flesh cannot. This enhancement through technology came initially in the form of the magnifying lens. Telescopes increased the reach and power of vision into the very far away. The planets were, for the first time, seen as other worlds – spheres rather than pinpoints of light.

Magnifying lenses brought the very small into view for the first time through the application of the microscope. In 1665, Robert Hooke published the book *Micrographia*, which contained intricate depictions of tiny animals and objects observed through a microscope. These included a flea (see Figure 6.1) and the cells within a slice of cork. This was the Royal Society's first ever publication, and it gave magnification center stage in the nascent scientific institution. The book also had great popular appeal, which meant important positive press for the burgeoning scientific movement. The general public was mesmerized by the images. The flea, known until then as a speck that bites, was shown to be a blood-sucking beast, with powerful jumping legs and a shiny armored surface. Microscopy was a kind of revelation for the scientific age.

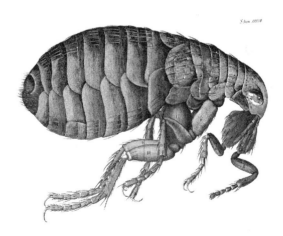

Figure 6.1 Robert Hooke's drawing of a flea viewed under a microscope

At the same time as magnifying technologies were overcoming the limits of human vision, technologies of *measurement* provided new ways of visualizing reality. For example, in 1714, Daniel Fahrenheit developed the first mercury-in-glass thermometer, which used the visual reference point of the expanding mercury against his famous scale, which set 100 degrees as core human body temperature, and 0 degrees as the point at which an equal mixture of water and salt freezes. This transformed the ethereal quantity of heat into a visual form.

Another scientific technique that brought data into a form that vision could analyze was mapping. A famous example of this is the work of John Snow, a physician who was the first to establish the cause of cholera outbreaks. The dominant theory during his time was that cholera was transported by miasma – a noxious form of air. Snow did not find this convincing and sought an alternative explanation. During a cholera outbreak in Soho in 1854, he recorded the locations that cholera cases were occurring and marked each one on a map. Over time he noticed a clear pattern. He found that the cases clustered around a water pump on Broad Street. Although he did not know what in the water could cause cholera, he managed to persuade the council to remove the water pump. The epidemic promptly dissipated. Later research subsequently found out that the well had been built only three feet from an old cesspit, which had been leaking fecal bacteria.[8] Snow's technique of mapping data into a visual pattern that the eyes and brain can interpret has become standard across countless forms of science now.

Another classic example of how mapping is used to make data visible for science is the research done on earthquakes. The theory of plate tectonics states that the Earth's crust is divided into plates that cause earthquakes when they collide or rub against one another. One line of supporting evidence is the pattern that emerges when major earthquakes that occur around the globe over a long period are plotted on a map of the world. They create a pattern that signifies the location of plate boundaries.[9] Through mapping techniques such as this, a set of dispersed or separated events can become integrated into visual evidence for a theory.

Exercise for empirical awareness: Getting to know trees

An empirical attitude involves attending closely to information from the senses, and interpreting and classifying that information based on a rational scheme. This involves being awake to nature and your environment. For this exercise, you will be going for a walk in a park or the countryside, preferably using a route that is familiar to yourself. For the walk, you will need a camera and a notebook. Your aim is to catalog all the species of tree that you encounter along the way. For many in the modern world, we live amongst trees but don't recognize them. Changing this situation is both empowering and a good exercise in empirical observation skills.

First, draw a map of the route and note the locations of the trees. For each tree that you don't know the species of, take photos and samples of leaves, draw a diagram of the overall shape and note down the approximate height.

When you get home, the aim is to identify the species you have recorded and find out information about each one, including their scientific Latin name (for example, an oak is *Quercus robur*). Find out if they have edible fruit or flowers, particular uses or medicinal attributes. You can use apps such as *Tree ID*, *PlantNet* and *MyGardenAnswers*, which use recognition systems to classify species from photos or feature descriptions. By the time you read this, there may be apps that work even better. Explore your options.

Even with the help of books and technology, the task is still hard, and often you will be unsure. Seek out more evidence and continue with the task until you find an answer. This kind of tenacity is as important to an empirical attitude as anything else. The ultimate aim of this activity is to develop the capacity to integrate your senses and your thinking, and so move beyond perception to knowledge-led observation.

New channels of scientific vision: Opening up the electromagnetic spectrum

Until the twentieth century, science was reliant on light and sound for information about the world and the cosmos. Although magnification technologies had increased the range of what scientists could see, it was still light passing through the instruments and providing the data. This was all to change with the discovery of the electromagnetic spectrum and harnessing it for scientific uses. Science now divides this electromagnetic spectrum into the following categories (starting with the highest frequency and finishing with the lowest): *Gamma rays, X-rays, ultraviolet waves, visible light, infrared waves, terahertz waves, microwaves and radio waves.* Visible light comprises just a small fraction of the spectrum, so there is a lot of additional information for scientists to tap into. Below I provide some examples of how science has used these different parts of the electromagnetic spectrum to open up greater access to the empirical universe.

Gamma rays have uses in medicine and physics. They can be used to build up a picture of what is going on inside a person's body, by giving a person a radioactive substance and then monitoring the emission of gamma rays. This can be used in medical diagnosis and in biological research. In physics, gamma rays are used to ascertain the presence of particular elements from a distance, because different elements emit different gamma ray signatures. In 2011, a probe called MESSENGER was sent to the planet Mercury. It used a gamma ray detector to measure gamma rays emitted by the surface of the planet, and from that information it could infer the presence of particular elements and minerals.

X-rays have various scientific uses, including in astronomy. The Chandra X-ray Observatory is a space observatory that monitors X-ray emissions from stars and planets. This provides information about the composition, temperature, and density of distant worlds. Also, X-rays were integral to the discovery of DNA. Rosalind Franklin used a technique called X-ray diffraction to study the structure of DNA. In this technique, X-rays are shone at a sample of a molecule, the X-rays are scattered and the resulting image is interpreted by an expert to infer the molecule's structure. Franklin's resulting image, Photo 51, was used by Watson and Crick to help develop their double-helix model of DNA.

Infrared radiation can reveal objects in the universe that cannot be seen by visible light. This is because their longer wavelength allows them to pass through regions of gas and dust in space with less scattering and absorption than visible light. The James Webb Space Telescope (JWST) has three infrared instruments, which are used to study the formation of galaxies, stars, and planets. It is scheduled for launch in 2018.

Microwaves are used in radar technology. The radar altimeter on the Ocean Surface Topography Mission (OSTM)/Jason-2 satellite can determine the height of the sea surface. It beams microwaves at two different frequencies (13.6 and 5.3 GHz) at the sea surface and measures the time it takes the pulses to return to the spacecraft. The pulse can determine the sea surface height to within just a few centimeters. Microwaves have also been used to measure the background radiation from the Big Bang, using microwave detectors on satellites (as previously mentioned in Chapter 3).

Radio waves are used by scientists to study space in various ways – they have the advantage that clouds and rain do not affect observations. By studying radio waves emitted from planets, stars, and galaxies, astronomers gather data about their structure and movement. Since radio waves have a low frequency, the collecting telescope must be very large. For example, the Parkes Radio Telescope has a dish 64 meters wide. China has just built the largest ever radio telescope, nicknamed "The Eye of Heaven" at a colossal 500 meters across. It is, at the time of writing, undergoing testing.

These scientific instruments that detect the different ranges of the electromagnetic spectrum have extended the reach of science beyond the senses, but there remains a challenge in representing that information in a way that the human body can access. All the information that scientific instruments gather must, in the end, be funneled back via the five senses into the nervous system, in order that the scientist may analyze that information using their brain. This is usually achieved by way of some visualized readout, like a medical X-ray photograph, which translates X-ray data into pictorial data. So the bodily senses remain an immovable funnel through which all scientific evidence must enter at some point for analysis and interpretation, yet they may lack the capacity to relay the whole of reality, even with the support of scientific instruments that deploy the full range of the electromagnetic spectrum.[10] The hope of science is that the parts of reality that are yet to be observed will come into view one day, but that is a promissory faith. The mystics, mediums and shamans who pursue the spiritual transcendent suggest that it is not in the sensory-empirical domain at all, but rather is accessed within – through altered states of consciousness.

Spirituality and the Transcendent

In the mid-1700s, the Swedish scientist Emanuel Swedenborg had reached the pinnacle of his career. He had written treatises on topics as diverse as cosmology, human perception, anatomy, algebra and chemistry, while devising cutting-edge engineering solutions for mining and dock construction. Then in his mid-50s, following a period of personal crisis, his mind turned away from science towards mysticism and religion. He subsequently wrote scores of works on these topics, while personally cultivating mystical practices for directly experiencing what he described as transcendental realms. He claimed to be able to move his consciousness between this world and others, and claimed that this capacity to experience transcendental realities was inherent to any human mind. He described it as opening the inner eye of spirit:

> A man cannot see angels with his bodily eyes, but only with the eyes of the spirit within him, because his spirit is in the spiritual world, and all things of the body are in the natural world... Moreover, as the bodily organ of sight, which is the eye, is too gross, as everyone knows, to see even the smaller things of nature except through magnifying glasses, still less can it see what is above the sphere of nature, as all things in the spiritual world are.[11]

Swedenborg spent countless hours in this contemplative state, immersed in visions. He found that the doorway to these inner worlds opened up when awake yet almost asleep. This state is now referred to as *hypnagogia*, and research has found that many people experience vivid visions in it.[12] Swedenborg's experiential pursuit of the transcendent has been a powerful influence on many important writers and mystics since him. The poet-artist William Blake credited Swedenborg with helping to understand his own visionary experiences, and incorporated his name in some of his etchings. Aldous Huxley quoted Swedenborg in his works on transcendental experience, and more recently Raymond Moody credited Swedenborg with helping him to make theoretical sense of his data on near-death experiences.[13]

Another important connection between Swedenborg and contemporary spirituality is William James, whose work *The Varieties of Religious Experience* is a key founding document for contemporary spirituality. What is not widely known is that James' father was a follower of Swedenborg, so his childhood had been steeped in beliefs and practices about knowing higher worlds. James created important links between rational philosophy and scientific psychology, while professing a mysticism that had a strong Swedenborgian flavor. For example, here is a passage by James that could easily have been written by Swedenborg:

The further limits of our being plunge, it seems to me, into an altogether other dimension of existence from the sensible and merely 'understandable' world. Name it the mystical region, or the supernatural region, which ever you choose. So far as our ideal impulses originate in this region... we belong to it in a more intimate sense than that in which we belong to the visible world, for we belong in the most intimate sense wherever our ideals belong. Yet the unseen region in question is not merely ideal, for it produces effects in this world.[14]

While James was giving renewed philosophical credibility to the idea of the transcendent, other contemporaries of his were giving it popular appeal as something that could be genuinely experienced. One was Evelyn Underhill, whose ideas were introduced at the beginning of the chapter. Her book on mysticism focused on the practical aspects of pursuing transcendental experience, and how it was something that both ran through and beyond religion. She wrote that transcendental experience is available to anyone to the extent that they are not overly caught up in the sensory and the empirical. In states of contemplation and ecstasy, when attention is less focused on observation and thought, the mind becomes capable of picking up messages from transcendental levels of inner reality.

Rudolf Steiner's anthroposophy movement provided another powerful force for opening up the transcendent to the modern seeker. Central to anthroposophy is the notion that higher worlds can be experienced in altered states of consciousness, and that the human being bridges all levels of reality, by way of "subtle bodies", including the etheric body and astral body.[15] To develop the capacity for inner silence and deep tranquility is, for Steiner, the foundation for becoming aware of these other layers of reality. Steiner's work was esoteric but had varied practical influences on culture, including the Steiner Education movement. There are currently around 1,000 Steiner schools (aka Waldorf Schools) in 60 countries around the world, which promote a holistic form of education that allows children to explore spirituality in a non-doctrinal way alongside their standard curriculum.

Another important influence on transcendental spirituality was Frederic Myers, who wrote *Human Personality and Its Survival of Bodily Death* in 1903. Myers presented the idea that all human beings have unconscious layers, which together comprise the *subliminal self*. He claimed that transcendental experiences stem from the subliminal level of mind. Myers conceived of the brain not as the creator of consciousness, but as the *transmitter* of consciousness, much like a radio or TV transmits signals. If a brain is a transmitter of consciousness, then psychoactive drugs or brain surgery change the transmitter and the proverbial TV signal. Science currently has no idea of the mechanism by which a brain could create the first-person field of consciousness from its physical workings, *nor* of how it could transmit it as a signal or field. The transmission theory has the benefit, in the context of transcendental experience, of not precluding experiences of alternative realities, for it would in principle be possible to unlink whatever mechanism keeps consciousness 'in' the brain.[16]

At the science-spirituality interface: Transpersonal psychology and shamanism

In the 1960s, a group of psychologists including Abraham Maslow, Stanislav Grof and Anthony Sutich set forth a manifesto for a new kind of psychology that drew on the ideas of the aforementioned pioneers – Swedenborg, James, Myers, Steiner, Jung and others – to bring spirituality into the scientific study of human beings. The movement was named transpersonal psychology, and it continues today as a key area of the research on experiences and phenomena that have transcendental, spiritual or numinous qualities. It theorizes about the transcendent in ways that are free of religious theology, and has provided a rich source of ideas and practices for the field of popular spirituality.[17]

Among the work done in transpersonal psychology on bridging science and spirituality, the research of transpersonal psychiatrist Stanislav Grof is an important example. He has developed a framework for understanding the transcendent and altered states of consciousness that integrates ideas from quantum physicist David Bohm.[18] Key to his theory is Bohm's distinction between the *implicate* and *explicate* levels of reality. The implicate order is a level of reality that sits outside of space and time and is used by Bohm to explain quantum phenomena like the wave function, quantum superposition and quantum entanglement. The explicate order is that which we experience in normal life; it is structured by space, time and the abstractions of thought.

Grof based his ideas for understanding human nature on this two-level foundation, suggesting that human beings have manifestations in both the implicate and explicate order. The explicate person is embodied in space and time, while the implicate layer of our being exists beyond space or time, and hence beyond the body. In line with Myers' transmission theory, Grof takes the position that the brain transmits and filters the field of consciousness, which stems from the implicate order. In this framework, transcendental experiences can be understood as a detuning of consciousness from the explicate order, and tuning into the timeless and eternal implicate order. The human mind can thus transcend space and time, and thus transcend individuality and the normal limitations of human life and the senses.[19]

Supporting the work of Grof, Charles Tart's work on altered states of consciousness has studied the conditions in which consciousness can make the jump into transpersonal states. Tart sees the physical world not as a container of mind of consciousness, but as a stable patterning of consciousness itself (in the line of Kantian and idealist philosophy). The stream of thought that runs through waking life is in his view a way of keeping this structure in place, and so disrupting it is the first step towards an altered state. Tart focuses mainly on meditation and psychedelic drugs as forms of consciousness alteration, while Grof developed his own method called holotropic breathwork, which combines deep breathing with visualization.

Another popular method for developing awareness of transpersonal state of consciousness is lucid dreaming. In a lucid dream, a person becomes conscious of the fact that he or she is dreaming and then consciously and voluntarily moves around the dream landscape and summons beings or helpers to the dream. This was originally practiced in Buddhism as *dream yoga*, and is widely considered in contemporary spirituality to be a way of providing access to transcendental levels of awareness, and to latent powers of creativity and healing.[20]

Exercise for transcendental awareness: How to learn lucid dreaming

In a lucid dream, the dreamer becomes aware of being in a dream, remains in it, and is able to consciously act within the dream space. Anyone can do lucid dreaming by committing to a training regime. Below is a set of five steps that you can follow to try. If you succeed, you will open up a new expanse of supersensory landscapes to your conscious mind, which you can use to overcome psychological fears and traumas, solve problems, tap into your creativity, or simply become more aware of the nature of realities beyond the physical.

The first step in this exercise is to keep a dream journal and write down whatever you can recall of your dreams in the morning. There are dream journal apps that one can use for the task. Your dream recall and intensity will increase if you maintain such a diary, and once you remember your dreams regularly, you are ready to try to experience lucidity in a dream.

The second step is to do *reality checking* while you are awake. To do this, every hour or so, ask yourself, "Am I dreaming?" and try to push two fingers on your right hand through the palm of your left hand. Then look around at the environment you are in, and ask if it is real or if you are imagining it. You can set hourly reminders on your phone to help with doing this regularly. You can then use the same reality checking injunction in the dream, and you will find that if you do, you will be able to push your fingers through your palm. If you want to become a lucid dreamer, you will need to make reality checking part of your life too.

As the third step, write down your aim to have a lucid

dream, and put the note somewhere prominent, so you will see it every day, or create a reminder or screensaver on your phone. This all helps to *incubate* the intention.

The fourth step is, before you fall asleep, to repeatedly say to yourself, *"I am lucid dreaming tonight."* You can even plan where you will be in your lucid dream – perhaps you have a favorite place. This helps to give the intention extra clarity and priority.

The fifth step, to introduce if you try the above for a week and do not get success, is the *'wake-back-to-bed' method*. Set your alarm for six hours after you fall asleep. After the alarm goes off, get out of bed and make yourself fully alert. Go and read a book about lucid dreaming, or read accounts of other people's lucid dreams on the many websites devoted to the topic. Stay awake for 15 minutes and go back to bed. Set your intention to lucid dream again, and relax into sleep.

Another important influence on transpersonal psychology was the scholar of comparative religion Huston Smith, whose philosophy was in the tradition of Neoplatonism. He argued that a tree-tier view of reality is common to all religious traditions. In his book *Forgotten Truth* he describes this threefold structure as follows: The *terrestrial* plane (the gross, the material, the sensible, the corporeal, the phenomenal); the *intermediate* (the realm of archetypes and spirits which governs the terrestrial plane, and can be experienced directly in altered states of consciousness); and the *celestial* (experienced as God, Unity, the Love beyond opposites, or Primal Light, and can only be experienced indirectly). Around and beyond all these three levels, there is the *Infinite* – an all-inclusive totality that is beyond and within everything and defies all categorizations and opposites.[21]

The belief that reality is multilayered and extends beyond the physical has been not only central to the religions that Huston Smith surveyed, but also to indigenous spiritual traditions that predate scriptural religion. Shamanism is the modern name given to spiritual practices and ideas that are thought to have their origins in the Neolithic period (20,000–2000 BC), and which are still practiced in many tribal cultures today. Some shamanic rituals may have been used continuously now for 12,000 years in some parts of the world, which makes them a candidate for the most enduring cultural practices in human history.[22]

Linked to the rise of transpersonal psychology, techniques from shamanism have seen considerable growth as a form of spirituality in the West since the 1960s, including the use of drumming and dancing as a way of altering consciousness, and the use of plant medicines.[23]

Shamanic practitioners train in how to enter into a state of expanded consciousness that allows access to transcendental levels of reality. The *core shamanism* method commences with setting an intention at the beginning of the journey to gain transcendental information to help or heal, and then imagining moving into a higher world, middle world or lower world, while listening to a repetitive drumbeat. As the journey progresses, the practitioner will experience themselves journeying to other dimensions, and interacting with beings or spirit animals that apparently exist there.

Following the ideas of Carl Jung, transpersonal psychology conceives of the unconscious as a parallel universe to waking consciousness that extends beyond the brain and body. While the conscious self is experienced as *me* and *mine*, the deeper layers of the unconscious become increasingly other and *not-me*. This helps to explain why the beings and realms encountered in the shamanic journey are experienced as beyond the self.[24] The philosopher Terence McKenna describes this as follows:

I believe that the best map we have of consciousness is the shamanic map. According to this viewpoint, the world has a 'center,' and when you go to the center – which is inside yourself – there is a vertical axis that allows you to travel up and down. There are celestial worlds, there are infernal worlds, there are paradisiacal worlds. These are the worlds that open up to us on our shamanic journeys, and I feel we have an obligation to explore these domains and pass on that information to others interested in mapping the psyche. At this time in our history, it's perhaps the most awe-inspiring journey anyone could hope to make.[25]

Shamanic experiences are, from a transpersonal perspective, those in which unconscious material that is normally confined to dreams comes directly into consciousness, thanks to the induction of the trance state. The landscapes and beings that emerge in shamanic experiences are sufficiently stable that one can return to them and meet the same beings on multiple occasions. Jung himself describes this in his own shamanic journeys and his encounters with spirit beings, such as a male character he calls Philemon, who he meets on many occasions and even dialogues with.[26]

In addition to the method of using drumming and chanting to bring about a visionary trance, shamanic methods that employ sacred plant medicines to initiate an altered state have enjoyed a renaissance over the past fifty years. Arguably the most famous account of a spiritual experience occasioned through the use of a shamanic plant medicine is Aldous Huxley's *The Doors of Perception*, in which he describes his experiences following the ingestion of mescaline. Mescaline is the active ingredient of the peyote cactus, which is used by shamans of Central America as their ceremonial sacrament. Huxley interpreted his experiences as an expansion of his consciousness into areas of 'Mind at Large' that are normally closed off from the senses. He came away from his experience with mescaline convinced of a greater transcendental reality that subsumes this one. He wrote:

> That which... is called 'this world' is the universe of reduced awareness, expressed, and, as it were, petrified by language. The various 'other worlds,' with which human beings erratically make contact are so many elements in the totality of the awareness belonging to Mind at Large... Like the earth a hundred years ago, our mind still has its darkest Africas, its unmapped Borneos and Amazonian basins. In relation to the fauna of these regions we are not yet zoologists, we are mere naturalists and collectors of specimens... Like the giraffe and the duck-billed platypus, the creatures inhabiting these remoter regions of the mind are exceedingly improbable. Nevertheless they exist, they are facts of observation; and as such, they cannot be ignored by anyone who is honestly trying to understand the world in which he lives.[27]

One of the most popular contemporary means for initiating a shamanic visionary experience is via rituals that center on the ingestion of ayahuasca, a ceremonial concoction made originally by tribes of the upper Amazonian region. Ayahuasca contains DMT, a psychedelic compound dubbed the "spirit molecule", for its capacity to bring about experience of alternate realities and deep insights. When it is taken in its raw form it leads to visionary out-of-body experiences that last just 10 minutes or so.[28] The addition of ayahuasca vine (containing MAO inhibitors) to the DMT-rich plant ingredient (chacruna) makes DMT orally active in ayahuasca, and the result is that the visionary state lasts hours rather than minutes.

Thousands of people every year head off to South America or the Netherlands to experience ayahuasca's visionary and healing properties, leading to an expansive industry of transcendental tourism.

There are many books and articles describing first-person experiences with ayahuasca (including one of my own – see endnote).[29,30] These reports invariably testify to highly structured and meaningful journeys, in which the person often experiences facing their own fears and deepest issues, while also coming into direct communication with beings of higher intelligence, whose concern appears to be love and healing.

In addition to this first-person case study literature, there is now also an expanding scientific corpus on psychedelics and their various effects on mind and body. The research within it employs systemic ways of collecting and analyzing data from multiple participants, and externalizes experiences using questionnaires or physiological measures. It has become an important and lively interface between science and spirituality. On the subject of ayahuasca, research has shown that taking it in a context of spiritual ritual has positive effects on depression.[31] Another research study supported the profoundly transcendental nature of the ayahuasca experience; in a sample of 30 people who had taken it, 47% reported encounters with 'suprahuman spiritual entities', and 37% experienced being in other universes and having encounters with their inhabitants.[32]

An independent scientific research study has recently been launched looking at the longitudinal effects on well-being of attending ayahuasca retreats at an institution called *The Temple of the Way of Light* in Peru. It is being conducted by ICEERS (International Center for Ethnobotanical Education Research & Service) in collaboration with the Beckley Foundation. All participants are asked to respond to a set of validated questionnaires measuring quality of life, mental health and well-being before and after the retreat, and then again at 3, 6 and 12-month intervals after it.[33] A preliminary analysis of data from the study, reported at the Psychedelic Science 2017 conference, showed a positive effect of ayahuasca on grief; 100% of those in the study who reported that they were grieving a loved one showing significant alleviation of their symptoms three months after the retreat, and indeed they surpassed the well-being scores of individuals who reported no mental health problems at the outset of the study.[34]

Using psychedelics as an aid to spiritual development has risk attached. It leads individuals into states of consciousness that mystics who pursue the contemplative path may have half a lifetime to prepare for. Ayahuasca and other shamanic plant medicines can occasion difficult and dark experiences as much as illuminating and celestial ones, and anyone motivated to explore the path of shamanic plant medicines should be armed with a clear understanding that such experiences are not particularly fun or recreational, and that it may take months or years to integrate the experiences into life fully. The spiritual context of any such journeys are integral to how they unfold, and in the end, any insights gained will only be fuel for personal change in the context of daily practices and ethical commitments.

It is also important to clarify that shamanic visions are markedly different from the hallucinations that are indicative of psychotic mental illness. Shamanic journeys bring fully immersive divulgence of worlds and realms that are experienced as more real, more intelligent, and more alive than this reality. They provide for insight and creativity, and can be used to help heal oneself and others. There are reports of individuals coming up with musical, mathematical or scientific breakthroughs in such states of consciousness,[35] and as a number of studies (including the aforementioned ICEERS study) have found, they can help heal emotional trauma.

In contrast, hallucinations that occur as symptoms of psychosis are associated with anxiety, depression and negatively charged thoughts. They lead to a decline in personal functioning, not to integration or developmental breakthroughs. While psychotic hallucinations are distressing, they are not necessarily meaningless. They can be considered as communications from the unconscious, much like intuitions or visions. The Maastricht Approach to voice hearing pioneered by psychiatrist Marius Romme theorizes that disturbing hallucinated voices represent traumatic events and associated emotions that have not been consciously recognized and worked through. Romme argues that hallucinated voices can be dialogued with to help identify an underlying emotional problem or crisis. Case examples of the Maastricht Approach and its therapeutic application are provided in the book *Living with Voices*,[36] as well as in the eloquent testimony of Dr. Eleanor Longden, who in a TED talk describes her journey from diagnosis of schizophrenia and hearing voices, through to being a world-renowned speaker and psychologist.[37] Longden describes how on one occasion during her student years, a male voice dictated an entire exam answer to her, which shows how coherent and informative hallucinated voices can be, even in the context of mental illness.

Some of the most intense and transformative transcendental visions that a person can have are those that occur in near-death experiences. It is to them and their scientific study that we turn now, as another important example of how empirical science and transcendental spirituality can helpfully overlap and interact.

The study of near-death experiences

In 1975 Dr. Raymond Moody published his book *Life After Life*, which coined the term *near-death experience* (NDE) to describe the vivid experiences reported by individuals who had temporarily died and then been resuscitated. It presents case studies of the NDE, and describes a common set of themes in them. Moody can't have guessed in his wildest speculations how well it would do – it has now sold over 13 million copies. It was clearly a topic that the spiritual zeitgeist was hungry for.

These core features of the NDE that Moody listed in the book are as follows: awareness of being dead; feelings of peace, euphoria and happiness; observing one's body from above;

entering a tunnel and moving towards a light; perceptions of heavenly or hellish landscapes; encounter with deceased relatives and religious figures; experiencing a life review; and the perception of aesthetic sound or music. These NDE features have been found to be fairly common across cultures.[38]

Over 600 articles have been written on NDEs since Moody. The consensus from large studies run at resuscitation units is that about 10–20 percent of people who are resuscitated after cardiac arrest have NDEs, or at least remember them. The presence and content of NDEs have been found not to be related to medication that might induce hallucination, for example, NDEs occur after severe road accidents in people who were not on medication.[39] NDEs occur in children as well as adults, including children who are too young to have a conception of death or the afterlife. Indeed, evidence suggests that a higher proportion of children have NDEs than adults.[40]

Research convincingly shows that NDEs lead to transformative developments in the lives of those who experience them. Longitudinal effects following an NDE persist for years; NDEers frequently find their interests drawn towards spiritual matters after the event.[41] Several studies have followed survivors of cardiac arrest and resuscitation over a lengthy period of time. Pim van Lommel and colleagues followed a group of NDE patients over a period of 8 years, and compared them with a matched group of cardiac survivors who had not had an NDE. For those who had had an NDE, even at 8 years there were notable effects: an increased interest in spirituality, an appreciation of ordinary things, and an increased empathy towards others.[42]

David Lorimer has researched the link between NDEs and ethical transformation. He proposes that the life review in NDE, which is experienced as a graphic overview of one's life in light of what one has done to bring love and happiness to others, leads many who have gone through an NDE to embrace an ethic of interconnectedness. This is an understanding that everything one does is part of an intricately connected fabric of life and consciousness, and so influences others in a multitude of ways. Adopting this ethic reflects in people changing their lives after an NDE towards a more prosocial or healing goal, and giving up on the motive to accrue wealth or achieve status.[43]

The various reported effects of NDEs do not by themselves attest to whether NDE is a genuine experience of the transcendent or some other kind of brain-induced event. After all, a brush with mortality is enough by itself to create profound change in a person's life. The academic community is split on how NDEs should be interpreted, with some adopting *in-brain* theories of the NDE, and others opting for *out-of-brain* explanations.

In-brain theories explain NDEs either as hallucinations caused by cerebral anoxia or a rush of endorphins.[44] In-brain theorists point to the fact that it is difficult to tell if the brain is fully shut down or not, so a continued level of awareness is possible even after cardiac arrest. Measures of zero electrical activity on the surface of the brain may hide ongoing electrical activity in lower centers of the brain. What in-brain theorists tend to gloss over is how a misfiring brain in a state of almost total shutdown would construct a complex and vivid experience that happens to coincide with what spiritual traditions have taught about death for millennia. That is quite a coincidence, and for some, it is too much to explain away physiologically.

Out-of-brain theorists argue that the NDE experience contains too many vivid elements to be explained by a dying brain. They point to the *intense sense of mental clarity and vividness* in most NDEs, which may be described as surpassing waking consciousness in clarity and sharpness. This is at odds with neurological conditions that should cause a fractured or diminished level of experience.[45] A 2011 review of in-brain and out-of-brain theories of the NDE, which was published in a mainstream psychology journal typically aligned to mechanistic science, concluded that researchers should remain open to both interpretations.[46]

Verbal testimonies from NDE experiences rely on the honesty and articulacy of the respondent, and this can always be open to question. One method that has tried to gain more objective evidence of the out-of-brain interpretation of NDE is the use of strategically placed images in resuscitation rooms on the tops of cupboards. This relates to the fact that those who have had an NDE report seeing their body from above during the episode, often hovering near the ceiling in the room. This means that a person could in principle, if such reports are veridical, be able to see things that are on the top of cupboards!

One major study that employed this method was the AWARE study by Sam Parnia and colleagues. The study took 4 years and involved 15 hospitals.[47] In order to assess the accuracy of any claims of visual awareness in NDE, each hospital installed shelves in resuscitation areas, with images placed on the top of them that were invisible to anyone who was not on a ladder. Unfortunately for the researchers, it transpired that 75 percent of resuscitations actually took place in locations without the shelves.

Nine percent of participants in the study reported NDEs. One man recalled precise details of what was going on in the room and who was there, and reported seeing all this from the upper corner of the room. However, no participants reported seeing, or recalling, any of the images on the shelves.

The findings of the AWARE study show just how difficult it is to gain empirical evidence of the transcendent in ways that would finally solve the mystery, even when a lot of resources are thrown at the task. In-brain theorists may suggest that this supports their view. The psychologist David Fontana, who took the out-of-brain interpretation, speculated that the whole question of the afterlife is perhaps supposed to be a mystery, and that the veil will never be lifted in its entirety for the matter to be put to rest.[48] He argues that if we had immediate, irrefutable evidence of the transcendent, and of life after death, this would obviate a key point of spirituality, which is to embark on a personal quest towards truth and wholeness, and to set out with courage into the mystery and see what one finds and brings back. This can only be done in the spirit of unknowing.

Some physicists have offered a prospective reconciliation between natural science and the study of NDEs by offering theoretical arguments for how the idea of consciousness beyond death can fit with physics. For example, cosmologist Bernard Carr has suggested that the human person may manifest not only in the three dimensions that we access in normal waking consciousness, but also at higher dimensional levels, and that these may not be reliant on the three-dimensional physical body. Modern physics already employs extra dimensions to explain the forces of nature. For example, the Kaluza-Klein theory explains electromagnetism by adding a fifth dimension to the four dimensions of relativity theory. Carr considers that this extra dimensionality could account for those paranormal and NDE phenomena that apparently defy the limits of three-dimensional space.[49] It is of relevance to Carr's theory that individuals who take DMT or ayahuasca frequently report experiencing more than three dimensions.[50]

Taking an approach based on quantum physics, physicist Roger Penrose has teamed up with anesthesiologist Stuart Hameroff to develop a theory that speculates on how consciousness could survive death. Quantum phenomena like electrons exist as superimposed waves, and hence contain possibilities that can coexist simultaneously (as famously illustrated in the thought experiment of Schrodinger's cat, where the cat is, due to "quantum superposition", both alive and dead simultaneously).

Penrose and Hameroff propose that consciousness may be linked to quantum effects in the microtubules (a network of tiny hollow tubes) in the brain's neurons. According to one widely accepted hypothesis called the "no-deleting theorem", quantum information is never destroyed and can be transferred to other physical entities or particles via the processes of quantum teleportation and quantum entanglement. If (a) this theorem is correct and (b) quantum information is indeed the mysterious substrate or correlate of consciousness, quantum information and consciousness would both continue to exist if the brain ceased to function. They would simply be transferred to entities or systems beyond the body, and hence would still be in existence in the universe somewhere. Using this logic, Penrose and Hameroff argue that consciousness could in principle continue to exist after physical death.[51]

In summary, the question of whether consciousness survives physical death remains open. It once looked like science would kill off the transcendent with a universe limited to observable objects structured in space and time, but that is long gone. The physics of higher dimensions and quantum phenomena has shown that time and space as we understand them *can* be transcended.

Chapter 7

Mechanism – Purpose

The 3rd of May 1621 was a day of political scandal in England. The trial of Sir Francis Bacon, one of the most powerful and respected men in the country, had ended with a guilty verdict. He had been adjudged to have taken political bribes and was sentenced to the Tower of London. The news was the talk of the taverns across the country, for Bacon was not only Lord Chancellor, but also a prolific author on matters of philosophy, science and law. He was something of a seventeenth-century celebrity, and his demise was a shock to all his supporters.

Bacon was released after just a few days in the Tower of London, on the order of King James I, but his political career was beyond saving. Knowing that he would not hold office again, Bacon devoted the rest of his life (which amounted to just five years) to furthering his vision for science and society. These ideas were presented in his book *New Atlantis*, which was published in 1627 – a year after his death. The book inspired the foundation of the Royal Society, and so became an important stimulus for modern science.

Central to Bacon's scheme was to move science away from matters of teleology and purpose, so that it could focus purely on mechanism and matter. In the language of Aristotle, which was the dominant philosophy of his age, this meant that science should not concern itself with final causes (the purpose or function of a thing), but instead it should focus on efficient causes (the process that gives rise to a thing), and material causes (the substance and structure of a thing).

Keeping final causes out of science was a challenge in Bacon's day, for the prevailing worldview was steeped in Aristotle and his view that *everything* has a purpose. Medieval theologians generally explained things according to God's purposes for man: *Why do we have eyes?* To behold the beauty of God's creation. *Why are there mountains?* So that man may climb closer to God. And so on. This fixation with final causes reached down to explaining individual lives too. For example, if a person was a slave or peasant, the explanation was that it was God's purpose for them, and hence the apparent injustices of society were explained away. If a person became ill, this was also God's purpose; potentially a punishment for a previous sin.

Bacon felt that science should unshackle itself from such speculations about purpose. He wrote: "The research into Final Causes, like a virgin dedicated to God, is barren and produces nothing."[1] He proposed instead that science should study nature solely in terms of matter and motion. Such a *mechanistic* approach would, he argued, lead to reliable and useful knowledge.

Despite wanting to eject the study of purposes from science, Bacon did not want to remove them from knowledge completely. He felt they had their place in the domain of metaphysics.[2] He wrote widely on matters of religion as well as science, and was closely associated with the radical Rosicrucian mystical movement.[3] The universe from the Rosicrucian perspective is imbued with intention and direction. The mechanistic world is a lower order of reality that supports the purposes of the higher order. One can equate this to how mechanistic human technologies support human purposes. The Rosicrucians used meditative techniques for cultivating an inner connection between the soul and the divine, from which one's higher destiny could be intuited. In summary, mechanism and purpose from Bacon's perspective are complementary partners in life and knowledge. Following his lead, in this chapter I explore how the distinction serves well as a basis for differentiating science and spirituality.

Science and the search for mechanisms in nature

Following ten years of space flight, in November 2014 the Philae lander of the Rosetta space probe touched down on a comet that was 317 million miles from Earth and traveling at 135,000 km/h. It was an astonishing achievement, even by the high standards of science. Images from the camera on the lander were beamed via the media into living rooms around the world, showing a dusty landscape in a far-flung corner of the solar system, henceforth forever connected to us by the instruments of science. Such feats speak of science's uncanny capacity for mechanistic prediction.

To see nature as mechanistic is to see it as the product of consistent forces and laws. It is to trust that nature does not do things in erratic ways, and that one can calculate to some accuracy what will happen in the future. It is this trust in nature's mechanistic workings that has led humans to invest billions in sending probes like Rosetta to distant planets and comets on the assumption that the spacecraft, years in the future, will arrive at exactly the right destination thanks to the predictable mechanisms of force, propulsion and gravity.

Mechanistic thinking involves reasoning about what causes change and motion in physical and organic systems, and seeks basic laws that govern causal interaction. This is reflected in the etymology of the word mechanism, which comes from the Greek word *mekhanos* meaning means or cause. Mechanics – the study of physical movement and change – has been a foundation of science ever since Bacon suggested that it should be, and Newton developed his theory of classical mechanics just over half a century later. This was added to in the early twentieth century by quantum mechanics, which studies how quantitative changes and effects occur at a subatomic level. Quantum mechanics requires a non-deterministic form of thinking, for there is uncertainty and non-classical motion built into its equations. However, it is still an essentially mechanistic approach to reality that studies the mathematical motion and interaction of subatomic particles and explains outcomes based on these, devoid of considerations of purpose.

A mechanistic explanation in science involves postulating a predictable set of causes that lead to an outcome or phenomenon. In physics and chemistry, a mechanistic explanation for a phenomenon is generally viewed as being a sufficient explanation for it. In other words, a mechanistic explanation answers the questions of how *and* why it comes about. For example, if you want to explain how (and thus why) a chunk of sodium moves about and flashes when it is put on water, you need to describe the mechanism that makes these effects happen. How it happens is that that sodium displaces hydrogen gas from water, and the plume of hydrogen gas moves the sodium around, while also leading to flashes as the hydrogen ignites due to the heat produced by the reaction. This answer provides a response to both how and why it happens, for in physics and chemistry finding a mechanism is usually considered a sufficient explanation. I will discuss later in the chapter how in the life sciences and psychology, the questions how and why tend to lead to different answers.

Exercise for mechanistic thinking: The mechanical interaction of sound and taste

Sound is caused by oscillating fluctuations in air pressure. Sound frequency is measured in hertz (Hz), which is the number of oscillation waves that pass a fixed point in one second. The human ear can hear frequencies approximately between 20 Hz and 20,000 Hz, although capacity to hear high frequencies reduces with age. Each note on the musical scale corresponds to a very specific frequency, for example, Middle C on the piano is 261.6 Hz. Frequencies double with each octave.

The exercise described here is designed to develop your awareness of how sound works from a mechanical point of view, and it relates to other senses such as taste. You will need to download a tone generator app to your smartphone or tablet. This can play tones at any frequency. You will also need to buy a bar of dark chocolate.

Research has found that listening to high frequencies makes food taste sweeter, while listening to low frequencies makes it taste more bitter.[4] This is referred to colloquially as 'sonic seasoning.' Eat the chocolate while having your tone generator play a high-frequency sound (e.g. 420 Hz), and then have a glass of water and eat another chunk while listening to a low-frequency sound (e.g. 85 Hz). Note if you experience a difference, and play around with the difference in frequencies to see if and when a difference in taste occurs.

Then try the task with several other people – tell them they are going to state whether one piece of chocolate is sweeter than the other, then give them the two pieces of chocolate while listening to the different frequencies, and see what they say. Does the predicted effect emerge? In my tests of the phenomenon with my wife and friends, it did. They were surprised that the two pieces of chocolate were actually the same. Whether or not you get the expected finding with this improvised experiment, you will in the process be developing your awareness of the mechanics of sound.

Mechanistic thinking in science and medicine tends towards *reductionism* – an approach to explanation in which things are explained by recourse to ever smaller processes and constituents. What the reductionist scientist finds upon examining matter is mechanisms-within-mechanisms, like a system of Russian dolls, each of which seems to make the bigger ones work.[5] For the reductionist, small things cause big things to happen – for example, little viruses and bacteria cause illness in bodies, and little subatomic particles cause light bulbs to illuminate. The reductionist tendencies of mechanistic science can be seen in how it depicts complex physical or biological processes.

An example of this can be seen with the process of photosynthesis. At a basic level, photosynthesis can be construed as the process within the chloroplasts of green plants that uses the energy of light to help transform carbon dioxide and water into glucose and oxygen. At the next level of complexity, photosynthesis can be divided into light-dependent reactions (by which electrons from light energy are used to create two molecules called ATP and NADPH), and light-independent reaction, also known as the Calvin Cycle (by which carbon dioxide and the molecules from the light-dependent reactions are converted into glucose). Each element of these mechanisms can be understood in mechanistic detail, along with the electron carriers that transport electrons between stages of the process.

At an even more microscopic level of detail, photosynthesis can be described through the language of quantum mechanics, which models the electrons that provide the energy for the process not only as particles but also as waves, and shows how interference and diffraction of such waves help energy move directly through the leaf to the reaction site.[6]

Advances in physical medicine are driven to a large degree by mechanistic thinking, and hence medicine tends towards reductionism too. To find the mechanism of a medical disease is to find an identifiable cause that can be removed or alleviated. This form of medicine has been successful in identifying how certain bacteria and viruses cause illness. For example, it was recently found that stomach ulcers are caused by an infection of the bacteria *Helicobacter pylori*.[7] This provides a mechanistic explanation for how an ulcer comes about, and immediately suggests ways of preventing and curing the problem if the bacteria can be eradicated or reduced. Surgery is perhaps the paragon of mechanistic medicine, and is an example of the success of this way of thinking to improve the quality of human life. However, mechanistic thinking does have abiding challenges and limitations too, which relate in no small part to its tendency to see all things as machines.

The machine metaphor in science

Mechanistic science has evolved in tandem with an understanding of the universe and living beings as machines. Astronomers of the early modern era described the universe as a machine that showed clockwork-like predictability. They believed that God had created the universe in the same way that humans make machines, and that this is why it is so mathematically predictable and elegant. Johannes Kepler (1571–1630), who discovered the elliptical orbit of planets, surmised that the solar system was a machine made of rotating crystalline spheres, which all moved in a kind of musical harmony. Figure 7.1 shows a drawing illustrating this from his work *Mysterium Cosmographicum*.[8]

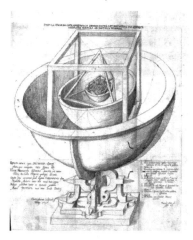

Figure 7.1 Kepler's machine depiction of the solar system

In addition to its use in physics, the machine metaphor was applied to the science of living things. Rene Descartes (1596–1650) described the body as a machine created by God into which the soul is inserted via the pineal gland in the center of the brain. In so doing, he implied that bodies are inanimate, and only the soul is conscious and alive. Animals don't have souls, so don't have a subjective field and can't feel pain. Having convinced himself that animals were machines with no soul, he had no problem in cutting them up for experimentation while they were still alive. Descartes had taken the machine metaphor too far, and had become blind to the reality of suffering in other species.

A century after Descartes, La Mettrie (1709–1751) took the soul out of humans too, and reduced them to just matter and mechanism, alongside the rest of the animal kingdom. He stated in his book *Homme Machine* (Man a Machine): "The human body is a machine which winds its own springs... the soul is but a principle of motion or a material and sensible part of the brain, which can be regarded, without fear of error, as the mainspring of the whole machine."[9] La Mettrie's human machine was without a divine designer – it was a product of blind nature and lacked any special status in the tree of life. He had laid down a key precursor to Darwinian theory and its mechanistic explanation of life, which emerged a century after him.

Following the lineage of Descartes and La Mettrie, contemporary biology is suffused with the machine metaphor. Richard Dawkins, in his influential work *The Selfish Gene*, refers to organisms as "survival machines – robot vehicles blindly programmed to preserve the selfish molecules known as genes."[10] In research on DNA, enzymes that facilitate the process of DNA replication are referred to collectively as "DNA replication machinery."[11] A recent article on how DNA synthesizes proteins refers to ribosomes as "the cell's protein-making machine."[12] Another recent article from the journal *Nature* on DNA repair refers to enzyme RecBCD as a "machine for mending DNA breaks."[13] In addition to this medley of machine metaphors, the cell is often referred to as a "factory" in which all these machines operate.[14]

The machine metaphor has taken center stage in psychology and cognitive science too. The metaphor that dominates at the moment is mind-as-computer. The language of computers gives mental phenomena a tangible, precise and mechanistic quality. Conceived of in this way, the mind runs programs and crashes when too many programs try to run at the same time. Thought is referred to as information processing, emotions are affective regulatory signals, visualizations are representations, short-term memory is a multi-store limited capacity system with an episodic buffer, creativity is divergent processing, while decision-making and the will are aspects of an executive function mechanism.

There are varied criticisms of this computation metaphor, including that computers have no subjectivity (as far as we know), no agency to make decisions themselves, and the language of computers reduces the rich inner life of the human being to a machine-like system of inputs, throughputs and outputs. Furthermore, as was discussed in Chapter 4, treating human beings or animals as machines has been frequently associated with diminished moral concern for their welfare.[15]

The machine-like predictability of nature is seen by many philosophers and scientists as evidence for asserting that the universe is without purpose. Mechanisms are sequences of events that are *pushed* by past conditions, rather than pulled towards future ones, and so the assumption is that a mechanistic universe is, by definition, not tending *towards* anything. The philosopher Bertrand Russell was one of the most forthright exponents of this view. He wrote that "man is the product of causes which had no prevision of the end they were achieving; that his origin, his growth, his hopes and fears, his loves and beliefs, are but the outcome of accidental collisions of atoms."[16] More recently Richard Dawkins has written that: "The universe that we observe has precisely the properties we should expect if there is, at bottom, no design, no purpose, no evil, no good, nothing but pitiless indifference."[17]

At the science-spirituality interface: The conundrum of purpose in physics and biology

While the mechanistic, mathematical and lawful qualities of physical nature are considered by Dawkins and Russell to be evidence for its non-living and purposeless quality, the very same qualities are considered by others as evidence for a divine mind that has designed the architecture of the universe according to elegant mathematics and reliable rules. Just as human technologies are mechanistic and consistent, precisely because they have been designed by a purposive intelligence, so the universe's many predictable mechanisms can be seen in this way.

The mathematical constants of nature are certainly set at precise values, and this has led some theorists to suggest that the universe appears to be *fine-tuned* for being stable and conducive to life. Martin Rees, an ex-president of the Royal Society, presents various examples of apparent fine-tuning in his book *Just Six Numbers*.[18] One of these is the astonishingly precise ratio of the strength of gravity (a very weak force) to the strength of electromagnetism (a strong force). The ratio is approximately 1: 1,000,000,000,000,000,000,000,000,000,000,000,000,000. This number allows atoms to hold together through electric charge and not be pulled apart by gravity. If the above number were shorter or longer by a few zeros, the universe could never have existed, claims Rees. From this and five other examples, he concludes that there *must* be an explanation for why such mathematical figures are as they are – it is not satisfactory to simply say that "it is so," and leave it at that.

Rees' favored explanation is that the universe is in fact just one universe amongst many, thus there are lots of opportunities to get one universe just right for stability and life, and we just happen to be in that universe. However, he does not dismiss the argument that the universe may be a product of higher purpose. He respects this as one way of making sense of the data, but chooses not to take that interpretive position himself. Meanwhile, there have been numerous physicists, such as John Polkinghorne, who do take fine-tuned laws as evidence for purpose in nature. A recent study showed that about 40 percent of physicists believe in God or a purposive Life Force.[19]

Another line of reasoning that implies a higher purpose to the universe is *simulation theory*. This theory states that the universe is a controlled simulation that is run by a species vastly more intelligent than we are. It may sound outlandish, but it now has many advocates, the most famous of whom is Elon Musk, scientific polymath and founder of PayPal, SpaceX and Tesla. Both Musk and another supporter of simulation theory, astrophysicist Neil deGrasse Tyson, have proposed the following argument for why we are in a simulation. On a timeline of millions of years, they say, one species across the trillions of star systems in our universe will become so advanced that it will become able to create a fully simulated reality. They point out that in just 40 years, human beings have gone from games like Pac-Man to hyper-realistic multi-player online games, virtual reality and augmented reality. Given millions of years and the likelihood of countless intelligent species across the cosmos, artificial universes are, they suggest, inevitable. If this is the case, there are likely a lot of simulated universes. If there is just one original 'base reality' and a large number of simulated realities, the probability is high that we live in one of the simulated ones, not the original. If that is the case, there is no logical reason to assume that 'base reality' would look anything like this one.[20]

Moving back down in scale from the cosmos to living organisms, the theory of evolution via natural selection has a curious and fraught relationship with the issue of purpose. When Charles Darwin devised his theory of natural selection, it was rightly seen as a triumph for mechanistic science. It showed how a simple causal mechanism could in principle explain the origin of life and the variation of species across the planet, without recourse to a divine purpose or plan. For those not familiar with Darwin's theory, the fundamentals are as follows: (1) When offspring are produced by organisms, the progeny may have genetic mutations and so show new physical and behavioral features; (2) In the struggle for life, some of these new features from genetic mutations make some individual organisms more likely to survive and reproduce, and some less so; (3) The organisms that are more likely to reproduce become more prevalent in subsequent generations; (4) Some of these successful variant organisms eventually become new species.

Although natural selection theory appears to be mechanistic, its method of explaining organismic features is actually *purposive*. For example, in answering why long-eared bats have long ears, why arctic foxes have white coats, or why giraffes have long necks, the explanation for these things according to evolutionary theory is the survival purpose that they serve. An example from human physiology helps to illustrate this. Scientists have long been trying to answer the question of why humans have an appendix. The appendix is a little tube that connects to the large intestine. Natural selection states that a feature such as this, which exists in all humans, must have developed initially as a result of mutations, and eventually become a standard feature because it serves a useful *purpose* for the organism as a whole, and so helps it to survive and reproduce. Determining *why* we have an appendix must, therefore, establish its purpose. This has been a mystery for centuries, but in recent years an explanation has been put forward. The purpose of it is, according to this new theory, to provide a safe haven for helpful gut bacteria, which can be deployed into the gut when illness flushes the good bacteria out of the intestine. So the appendix, it appears, is like a built-in probiotic drink. If this theory continues to be tested and supported, then the mystery of the appendix is solved – we would know its purpose, thus why it has emerged from evolution.[21]

The fact that evolutionary adaptations like the appendix serve a purpose has been interpreted in many ways. Richard Dawkins argues that such apparent purposes are side effects of a blind process operating over millions of years – he calls this archeo-purpose.[22] Another interpretation is the Intelligent Design argument, which claims that evolution is the product of intelligence that has purposively guided the development of matter into living forms and higher levels of complexity through the evolutionary process. Those who argue for Intelligent Design are mainly Christians who want to reconcile science with a designer God who is external to creation. While their argument is very different to Dawkins' atheistic Darwinism, they share a key feature in common; they both view the universe as lacking its own inherent purposes.

There are other ways of interpreting the evolution of life over time that require neither an external intelligent designer nor an entirely purposeless process. These alternatives honor the gradual mechanisms of natural selection and emergence of new species over billions of years, while proposing that some vital agency or purpose is inherent to the universe and to all living things. The guiding assumption that underlies these alternative views is that the universe is a living organism, rather than a machine, and thus has its own internal, vital creativity. This view of evolution – termed *organicism* – has a rich heritage.[23] It was propounded by the philosopher Henri Bergson, who suggested that evolution is motivated by an *élan vital*, which is a driving impulse to creativity in all things and the cosmos as a whole.[24]

Alfred Russel Wallace, who was the co-founder of the theory

of natural selection with Darwin, took a similar view. While Darwin retained a broadly materialist view of evolution, Wallace developed a view of the universe as imbued with a creative and directive force that he called the Life Principle, which drives evolution and biological growth in organisms. He wrote about this in a rarely cited book entitled *The World of Life: A Manifestation of Creative Power, Directive Mind and Ultimate Purpose*.[25] It was something of an embarrassment to the scientific establishment that such an eminent figure had come to this conclusion based on his life's research into natural selection.

Following Wallace's ideas about the role of a purposive force in directing growth, the biologist Hans Driesch conducted experiments on sea urchin embryos and found that large pieces could be removed, or cells moved above and replaced in different locations, without affecting the resulting embryo. It was as though the growth of the urchin was being drawn *towards* its ideal form and could flexibly re-orientate towards this form, despite considerable interference in the unfolding mechanism. Driesch concluded that development must have a purposive thrust, which he referred to as an organism's *entelechy*.[26]

The capacity of biological development to re-orient itself despite setbacks is striking in all species. In a particularly extreme human case of developmental flexibility, published in 2007, a 44-year-old man was found to have a condition that had caused almost his entire brain cavity to fill with fluid, leaving a brain that was less than 20% of normal volume.[27] Somehow over the many years that he had the condition, his brain had developed into a completely different, far smaller, structure than the normal brain. Yet it was working well enough for him to have a white-collar job as a civil servant and be a father of two.

Any explanation of biological development has to account for this capacity of development to adapt when it is disrupted, so as to continue development in the right direction towards the mature form. This is very difficult to account for mechanistically, particularly when you consider that human beings have just 20,000 genes (which is fewer than much less complex organisms, including water fleas), and each gene codes for just *one* structural protein. However, to build a human organism, information is needed not only for building the millions of proteins from which cells are made, but also for (a) how to build and place trillions of cells, each of which must be built as a specific type (e.g. nerve cell, muscle cell or hair cell) and put in a very specific location, (b) how to combine cells into organ systems, (c) coding innate instincts like suckling behaviors, (d) the blueprint for the whole structure of the organism, including all the different forms between baby and adult; and (e) instructions on how to adapt any aspect of biological growth in the face of disruption or damage. Can DNA really do all this, using a purely mechanistic process? If it can, how it does it is still a mystery.

Contemporary biologist Rupert Sheldrake has proposed that there is an additional source of information beyond genetics that informs development, which helps to account for its capacity to stay on track despite any disruptions. His theory states that all living things have a *morphic field* that defines development's final form, and the physical organism is drawn *towards* this over time. The structure of a morphic field is not fixed, but rather changes over long periods of time, acquiring as it does so a memory of the past. So in Sheldrake's theory,

Figure 7.2 The perfect edges of a tree's form: Evidence of a morphic field?

development is both mechanistic *and* purposive. It is pushed forward by the genetic control of protein synthesis, but is also *pulled* towards a final form that the morphic field depicts.[28]

The development of a tree can be seen as showing the functioning of a morphic field. A tree will develop a unique pattern of branches that is entirely different from its parents, and yet somehow ends up with the tip of each branch (neither the shape nor size of which has been genetically programmed) in a perfectly aligned form – see Figure 7.2 as an example. The question of how each branch of the tree knows to stop growing at the precise point to fit this overall form can't just be a matter of coordinating the duration of growth, because the unique shape of each branch requires growth for different durations, and the size of the tree may differ entirely from the parent trees. Sheldrake asserts that the tree's overall harmonious patterning, achieved despite no predetermined branch layout, is due to the structuring effects of a morphic field.

The theory of morphic fields also has potential application in understanding how animals can behave in purposive ways that defy standard explanations. An example of such a behavioral pattern is the Monarch Butterfly's migration from the northern states of the USA to Mexico. Each year millions of Monarchs, each just 0.75 gram in weight, undertake this journey of 4,000 miles. They all end up within a small patch of forest in Michoacán, Mexico, about 20 kilometers across. The trees in this area disappear under a carpet of butterflies when the Monarchs arrive. On the way, if the butterflies get pushed off course by weather or error, they change their course to return towards their goal, showing clear evidence of being oriented towards a goal. After mating in Mexico, the butterflies die before they can make the full return journey, but somehow their offspring, after metamorphosing from caterpillars into butterflies, know to migrate north. Hence the migration cycle completes over multiple generations.

A significant amount is now known about the mechanisms by which Monarch butterflies use information from the sun and their biological clock to navigate.[29] However, the purposive achievement of their migration is still a mystery, for no one knows how each tiny butterfly, with a brain the size of a pinhead, finds a precise goal thousands of miles away without any guidance from others or prior learning, and how alternating generations know to fly north or south.

Sheldrake proposes that migration behaviors such as this are evidence of a behavioral form of morphic fields. Behavioral fields are purposive, like developmental fields, in that they draw action *towards* an end point. They are not fixed but are laid down over time, as more and more individuals within a species enact a behavior.[30] A behavioral morphic field acts as a communal memory, so for Monarch butterflies, when they fly to Mexico and back again, they are using the species' communal memory of the journey's direction and purpose. With each successful journey, the realization of future journeys is made more likely as the field becomes clearer and stronger – a process called morphic resonance.[31]

In summary, the debates about whether there is evidence of purpose in the laws of nature, in evolution, or in biological development, lead to some of the most heated disputes in science. Any suggestion of there being evidence for higher purpose brings science up to and over the boundary that Bacon said it should not cross. But Bacon's injunction may be past its sell-by date; there are lingering mysteries in the natural world that warrant the critical consideration of whether nature is inherently purposive *and* mechanistic. We already understand the human being to work this way, so we have a model for how purpose and mechanism can co-exist peacefully.

On human purpose, free will and consciousness

In matters of human behavior, the reality and causal power of purpose is more acceptable to science. This is because it can be explained as the product of the brain's computing, rather than the manifestation of some invisible force. It is only mind and purpose *beyond* the brain that mainstream science mostly considers taboo. Hence, within psychology, mechanistic and purposive explanations are almost universally seen as complementary.[32]

For the researcher interested in explaining behavior, asking *how* leads to answers based on mechanisms, and asking *why* leads to answers based on purposes (also called goals, aims or intentions by psychologists). For example, the question "*how* does a person drive a car?" leads to answers about the mechanistic processes of handling pedals, gear lever and steering wheel, coordinating vision and action, and so on. On the other hand, the question "*why* does a person drive a car?" leads to answers about the purpose and goal-directed benefits of driving.

The fact of having purposes and goals is intricately linked to attention, awareness and consciousness. Goals are organized in a hierarchy of importance, and a person's conscious attention is usually directed towards their highest priority goal. Usually, this is a goal that is posing a *problem* and so needs attention directed at solving it. For example, when we start learning to drive, the whole activity is problematic, and so we need to think consciously about every movement, but eventually, we can drive without paying any conscious attention to it at all, so an experienced driver may even daydream while driving. Based on the link between goals, problems and consciousness, psychologist Albert Bandura suggests that the function of consciousness might be handling problematic purposive activity.[34]

When attempting to establish which goal is the highest priority to pursue at any one time, conscious attention will divert to the matter of goal selection, rather than goal pursuit.[33] Adult human beings can think through the future implications of pursuing different goals, and then select one according to the outcomes they have foreseen. It is this capacity for purposive foresight that is the basis in law for assuming that an adult has free will and culpability for his/her actions. For example in matters of criminal law, to be judged as guilty of a crime, a person must both have performed the criminal act, and be seen to have done it on purpose. For the criterion of being on purpose to be satisfied, there must be evidence of having selected and pursued an unlawful goal despite understanding its future consequences for harm. Society views children as less mentally able than adults to consider the future implications of actions when selecting goals, and therefore as possessing less free will and less capacity for autonomous, deliberate action. It is for this reason that the law prohibits children from goal-directed activities that have a high potential for harm, such as drinking alcohol, gambling, and driving.

Psychology, being a boundary discipline that balances the objectivity of science with the subjectivity of mental life, shades into spirituality in many places. I have looked at some of these areas in previous chapters, for example, transpersonal psychology and mindfulness studies. An area of overlap that is salient for this chapter is the study of *calling or vocation*. Research has shown that the majority of people experience being drawn towards a life purpose or job, in a way that feels like a "transcendent summons."[35] Religion in the West has typically interpreted that as the voice of God, but spirituality has a number of other ways of making sense of it. Indeed, when the term spirituality was given its modern meaning back in 1908, it was in relation to the personal feeling of higher purpose and vocation. Let's pick the story up there.

Spirituality, calling and vocation

In 1905, an American philosopher called Felix Adler wrote a short book called *The Essentials of Spirituality*. It was possibly the first text ever to use the word spirituality in the way that it is used now – to refer to something that extends beyond religion, as an inherent quality of being human. Adler described how spirituality necessitates a purposive orientation to life, and this does not require adherence to religious doctrine. The spiritual person thoroughly devotes his or her life to a higher moral calling, and feels enthusiastically drawn towards their ideal future. This is conveyed by Adler using the following metaphor:

> If a river had a consciousness like the human consciousness, we might imagine that it hears the murmur of the distant sea from the very moment when it leaves its source, and that the murmur grows clearer and clearer as the river flows on its way, welcoming every tributary it receives as adding to the volume which it will contribute to the sea, rejoicing at every turn and bend in its long course that brings it nearer to its goal. Such is the consciousness of the spiritually minded person.[36]

Adler's idea of equating spirituality with purpose had a strong modern lineage prior to the writing of his work. The German idealists Schelling and Hegel both developed secular yet spiritual philosophies that had purpose at the center. Schelling conceived of Nature as infused with a purposive and vital spirit, which over time organizes matter in ways that lead towards unity and consciousness. In his philosophy, as humans become aware of ourselves and our reality, they allow the universe to become aware of itself, thus participating in the process of divine self-realization through evolution. The divinity that Schelling describes is immanent and purposive, emerging out of chaos and being part of an open-ended evolutionary process, rather than setting the agenda from the outside:

> Has creation a final goal? And if so, why was it not reached at once? Why was the consummation not realized from the beginning? To these questions there is but one answer:

Because God is Life, and not merely Being. All life has a destiny, and is subject to suffering and to becoming. To this, then, God has of his own free will subjected himself.[37]

Like Schelling, Hegel spoke of an indwelling purposive Spirit that draws physical and mental reality towards wholeness and unity over time. He added in a dialectical quality to purpose, in the idea that Spirit evolves towards unity via *integrating opposites*. Humans can take an active part in this by bringing opposites together in schemes of knowledge. By doing this, they facilitate the development of Spirit to higher forms. As this process unfolds, human beings gradually wake up out of unconscious adherence to external rules, and become more self-reflective and conscious of their own responsibility to help create a better future.[38] This process of waking up to reality is an onerous one, for once we awake to our true nature as evolutionary agents, we understand that we are responsible for co-creating the future. This is a big responsibility, and there is an omnipresent urge in the person who has awoken to go back to sleep and leave the burden to someone else.

The ideas of Hegel and Schelling have been key influences on some of the most popular theories of contemporary spirituality. Ken Wilber's body of work on spirituality and science rests explicitly on the ideas of Schelling and Hegel.[39] For Wilber, the universe is an expression of Spirit, and evolution over time is directed towards increasing complexity. As higher states of complexity emerge over time, earlier levels of development are modified and retained within the embrace of the higher level. Wilber refers to this embracing of lower levels within higher levels over the course of evolution as a *holarchy*, and suggests that both matter and ideas are structured in this way. As an example in the domain of ideas, morality starts out as family-centric, then moves up to tribe-centric, then to nation-centric, and eventually to world-centric, each time reaching out more widely than previous levels while also including them. Religion and spirituality also evolve through grades of complexity, from tribal, to authoritarian, to individualistic, to relativistic, to systemic, and beyond. Each level must incorporate earlier expressions rather than rejecting them, to create the holarchy. The result over time is a movement towards wholeness that encompasses all things.

At the level of the individual human, Schelling and Hegel describe finding one's purpose not as dutifully following a destiny or fortune that is dictated from on high, but as the effortful endeavor to find a personal direction that accords with the cosmos' own purposive drift towards unity. Vocation evolves over the course of a whole life, and must constantly be intuited and re-intuited over time. James Hillman has explored this idea in his book *The Soul's Code*.[40] Vocation, in Hillman's view, is the filtering down of purposive information from the archetypal layer of reality, beyond time, into the conscious mind. Like a magnet draws iron towards it, the soul draws development towards it over time, seeking expression in action and living. However, this is not a simple linear path, and a person's vocation may change over the course of life as self-understanding develops. A way of symbolically rendering this lifelong struggle to find and achieve a purpose is the *hero's journey* narrative, which underpins many popular stories.

Exercise for clarifying one's life purpose: The 12-months-left list

Recent research by Bronnie Ware found that many individuals who are close to death due to a terminal illness have regrets that in turn illuminate life's meaning and purpose.[41] She found that the top five regrets were: *I wish I'd had the courage to live a life true to myself, not the life others*

---

expected of me; I wish I hadn't worked so hard; I wish I'd had the courage to express my feelings; I wish I had stayed in touch with my friends; I wish that I had let myself be happier. From this, we may assume that the meaning of life, at least as it is perceived near the end, is about warm friendships, being authentic and open, experiencing happiness and joy, and not giving too much of one's waking life over to work.

In this exercise, you will reflect on what you would do with your life if you found that you had a limited time left. First, you need to get relaxed and present in the moment. Make sure you won't be disturbed. Then take a pen and paper, and imagine as vividly as you can that you have been told you have 12 months to live. You need to make a list of things that you would do in those 12 months, and who you would like to spend time with. Once you have your list, you need to write underneath each one why you aren't doing it now. See what you write by way of obstacles and consider whether these are obstacles that are preventing you from pursuing your life purpose now.

You may expect this would be a depressing exercise, but in fact it often brings an invigorating, life-affirming reminder of what really matters in life.

Spirituality and the hero quest

The hero's journey is a storyline that was originally defined by the mythologist Joseph Campbell.[42] Campbell studied myths and religious stories across the world, and found a common basic plot to many of them. In this plot, the hero starts in the ordinary world, perhaps as an outsider or unexceptional person or child, but then feels the pull of vocation drawing them to a bigger path or destiny. They are called to an adventure or quest, but initially refuse the call on the basis of it being too much to contemplate. Upon eventually embarking on the journey, or being involuntarily thrust into it, they encounter a mentor who provides guidance on how to move forward. They are then initiated into a new and unfamiliar world, at which point they undergo a series of tests and trials that risk death, all the while feeling animated by a dream or vision of the future that makes all the crises and sufferings worthwhile. Within an ultimate ordeal, they experience a vision of fear and darkness, representing their own shadow and capacity for evil, which they overcome. After taking the road back to the ordinary world, they return to society with new wisdom, and bring some great prize or elixir that heals the sufferings of the ordinary world.

This narrative structure is at the core of many enduring spiritual myths of East and West – including the ancient Greek myth of Jason and the Argonauts, the stories of Moses and Jesus in the Bible, the story of Arjuna in the Bhagavad Gita, and the journey of Siddhartha Gautama in becoming the Buddha.

Campbell's hero myth concept became an important influence on contemporary storytelling from the 1970s onward, as the first *Star Wars* films were created by George Lucas with the intention of crafting a science fiction film around Campbell's mythic structure. The films are based on an animist kind of spirituality; Luke learns the way of the mystical Jedi sect, and learns how to connect with the "force", an indwelling living power that connects the universe together.

Disney also latched on to Campbell's hero quest in the early 1990s, producing a short guide for its writers on how to create the perfect hero myth narrative.[43] This led to the plot of the 1994 film *The Lion King*, in which the lion king's son Simba is cast of out of the pride, and goes on a long journey through which he learns his true identity and returns to save the tribe. There are hundreds of other examples of this ancient spiritual story of the questing hero being co-opted into modern cinema, including *The Lord of the Rings*, *The Matrix*, *Avatar*, *Harry Potter*, *Final Fantasy* and *Moana*. Many of these films express an explicit non-religious spirituality that relates to the importance of courageously following a higher calling in the service of one's community and environment.

The hero narrative in myth and fiction is enduring and culturally universal because it reflects in symbolic and simplified form the quests that people undertake in the pursuit of achieving a challenging moral purpose or goal.[44] Anthropologist Paul Heelas has looked at the path of modern spirituality in relation to the hero quest, and finds a clear resonance between the two.[45] For the spiritual seeker, their quest typically starts with a feeling of brokenness and crisis. The initiating crisis may be painful, but it is a gift insofar as it acts as a spur on the path towards renewed meaning. The resulting quest transmutes episodes of suffering into meaningful steps or trials on a path towards a better world, and in so doing gives it a meaningful quality that makes it more endurable.[46] The quest leads through trials and challenges, including encountering one's own inner demons, towards renewed hope.

All the time, the lower self or *ego* is continually trying to sabotage this process. It feeds on fear, insecurity, division and greed, and may represent internalized cultural conventions that are based on fear and control. In order to move past ego, spiritual and therapeutic practices are called into service, and over time the seeker becomes more giving, more loving and more peaceful, despite the turbulence of normal life.[47] Eventually a breakthrough realization occurs that a higher spiritual presence (an oversoul / true self / God / life force / inner power / spirit guide) is directly available and can be merged with or related to in a way that allows the quest to reach some kind of fruition. Self-realization is not found in rejecting the ego, but by integrating it within the embrace of one's higher self, which in turn requires committing one's life to the service of others.

169

Astrology – Archetypal guidance or lingering superstition?

Among the tools and practices that spiritual seekers use as guidance on their quest, divination techniques such as astrology, tarot, dowsing and palmistry continue to have popularity. Astrology remains particularly prominent due the ubiquitous horoscopes in newspapers and magazines. While many spiritual seekers and scientists alike may consider astrology to be a lingering superstition, and wouldn't go near a magazine horoscope, there are arguments for why astrology can be practiced and interpreted as a means for exploring one's life purpose in ways that are congruent with science.[48]

Astrology is essentially a purposive tool. Where there are no purposes there are no problems, and where there are no problems, there is no need for guidance, whether divinatory or otherwise. Critics argue using astrology to explore one's purposive path into the future is an example of lingering magic and superstition in spirituality, but Carl Jung argued that it has its place if construed correctly. He stated that if a person has lost touch with their own intuition and a conscious sense of their inner life (for example if they can't remember their dreams), they can use divination tools to bring their inner conflicts and challenges into pictorial or symbolic form, and hence into open discussion in an interpersonal context.[49] Astrology uses archetypal symbols and images that have a natural affinity with dreams and the unconscious, ambiguous statements that can be interpreted in a multitude of ways, and together these provide a canvas for a person to externalize the inner dramas that they may have been struggling to articulate. Jung was equally critical of literal interpretations of astrology. An unimaginative or literal mind, he said, will find nothing, and a superstitious or gullible mind will likely infer too much.[50]

Since Jung, a number of researchers have studied the links between star signs, personality traits and accomplishments, such as the link between the astrological location of Mars at birth and athletic achievement.[51] This body of research was enough to convince the famous psychologist Hans Eysenck that astrology had some validity, and he subsequently wrote of his support for it.[52]

The empirical research on astrology has been recently compiled and reviewed in the book *Cosmos and Psyche*, by Richard Tarnas.[53] Tarnas has further developed the Jungian approach to astrology. Like Jung, he is critical of literal interpretation of horoscopes or any other simplistic astrological predictions, but he believes that a nuanced understanding of astrology in relation to cultural history and individual life stories fits with a worldview in which the universe is seen not as an inert, lifeless thing, but one that has a conscious, purposeful presence. Following Pythagoras and Plato, he contends that the same archetypal and mathematical principles that structure the universe as a whole structure the unconscious and human lives, and hence one finds clear structural parallels between the cosmic and human levels. Planetary positions thus do not *influence* human fortune in Tarnas' view; rather they *correspond* to it in ways that reflect the different levels at which the same archetypal principles operate.

Further to this archetypal interpretation of astrology, there are arguments for why it may be congruent with modern physics. Einstein's theory of relativity states that space and time are not absolute but relative to the observer. In the theory, space contracts as speed increases, to the point that space reduces to zero at the speed of light. In other words, if we could catch a ride on a photon traveling at the speed of light, *there would be no time or space* – everything would exist in a mysterious singularity. Thus the apparent distance of the planets from us is a product of our perception and the speed we are moving at. The planets experienced as spheres hanging in the void of space and changing slowly over time do not exist in an absolute sense, but are really aspects of human experience and our particular perception of time and space. The fundamentals of astrological calculations – the relative positions of planets relative to the human viewpoint – are also clearly a product of the human mind. So we can say with scientific accuracy that the planets exist beyond consciousness in some unknown form that is not truly separate from us, or anything else. The ones we see are really within consciousness.

As for time – the dimension in which purpose and questing unravel – both relativity theory and quantum theory state that the future already exists in some sense, and that time travel is theoretically possible.[54] Divination tools like astrology may help to access levels of consciousness that connect to a reality beyond time, draw insights about the future down to the time-bound conscious ego, and so aid the faculty of intuition in determining where best to direct one's life.

Although astrology may have a valid rationale and a body of (contested) evidence to support it, there is no doubt that the whole domain of divination is riddled with charlatanry, from the comically ambiguous horoscopes in newspapers to the woolly advice offered via phone psychics and self-proclaimed mediums. Discernment and caution are crucial in exploring such matters. Nevertheless, to dialogue openly with someone about the archetypal motifs of the zodiac or the tarot *does* seem to help transform hazy inner hunches about one's future into a more explicit and external purpose. Divination is, for at least that reason, here to stay.

Chapter 8

Verbal – Ineffable

In 1919, a young soldier called Ludwig Wittgenstein returned from military service in the Austro-Hungarian army, after nine months in an Italian prisoner-of-war camp. He was physically frail, suffering from depression, and talking incessantly about suicide. Before the war, he had been an outstanding philosophy student at Cambridge and was a favorite of his lecturer Bertrand Russell. However, during the war he had found himself drawn more towards religion than philosophy, and had carried Tolstoy's *The Gospels in Brief* around with him everywhere, recommending it to his fellow soldiers. In 1918, during a short break from the front line, he had written a book called *Der Satz* (The Proposition), but later renamed *Tractatus Logico-Philosophicus*. It contained a series of pithy statements that Wittgenstein considered to be self-evident truths about the nature of language, thought and knowledge. He had sent it to two publishers but had been rejected by both. This had contributed to his depression a year later, but he continued to work on the book through his period of melancholy and crisis, and the revised version was published in 1921 to international acclaim. It would become one of the most influential books of twentieth-century philosophy.

Despite his success in the discipline, Wittgenstein was modest about what philosophy and science could achieve, believing both to be fundamentally limited by language. As a person, Wittgenstein embodied both the rational and the mystical. He had periods out of academia as a school teacher, and at one point he enquired about becoming a monk, but instead of taking holy orders ended up working as a gardener at the monastery. For him, rational knowing is facilitated and limited by language, while mystical knowing is ineffable, in that it transcends the capacity of language to convey. Towards the end of the *Tractatus* he wrote of this as follows:

> Whereof one cannot speak, thereof one must be silent... There is indeed the inexpressible. This shows itself; it is the mystical.[1]

Following Wittgenstein's lead, this chapter explores the verbal-ineffable distinction as a basis for distinguishing science and spirituality. I look at science's reliance on the language of mathematics and also its use of rhetorical language, before looking at the silent practices of mystical spirituality and its quest to find a knowing beyond words.

Science and its reliance on language

No matter whether a scientist gathers their data via a rover on Mars, a particle accelerator, an observational study of meerkats in the Kalahari, or a brain scanner, they are obliged to condense the information they gather into a fundamentally similar written form – the journal article. Across all branches of science, journal articles are typically fairly brief and have the same basic form: an introduction, aims and hypotheses, then methods and results and then an interpretive discussion. It is the article that formally enters the corpus of scientific knowledge, and from which facts and findings will be cited. So while scientific data do take many forms, scientific *facts* are by their nature linguistic.[2] The quantum physicist Niels Bohr realized this when he said that science is "suspended in language in such a way that we cannot say what is up and what is down."[3]

Given this reliance on language, the precision and clarity of communication is crucial, and this explains why concise, dry prose is much valued in science. The phrasing of scientific reports is such that all allusion, rhetoric, metaphor and unnecessary flounce is removed, leaving the bare necessities of what is needed to convey the point, even if that makes for unexciting reading. By making ideas as explicit as possible in clear writing, science aims to use language in as transparent a way as possible. However, for some tasks in their work, a scientist has no choice but to step outside scientific style, and use a more rhetorical form of language.

Scientists need money, for science is an expensive business. Philosophers can sit in an armchair and ponder for little cost, but scientists need to go places to collect data, costly instruments and a team of researchers to help out. So most scientists who want to do research must first apply for funding. This means convincing a charity, company or governmental organization to provide them with a research grant. A funding proposal will be submitted, in which the study is described along with arguments for why the research would be important, impactful and innovative.

A funding proposal will intentionally list all the possible upsides and positives of the study while minimizing the negatives or limitations. The scientist has to leave behind his or her critical reserve to engage in this process of persuasion. Funding bodies are generally concerned with real-world benefits of any research and not much with the advance of knowledge *per se*, so the researcher must find a way of persuading the funders that their study will have positive consequences for the betterment of society, even if the aim is actually just the advancement of knowledge.

Greg Myers, a professor of linguistics, studies how funding applications are written in biology, and has found that two elements contribute to the success of proposals. The first is the projection of *status* in the document, represented by the prestige of the scientist's university, the inclusion of famous scientists on the author list of the grant application, demonstration of previous funding and a track record of publications. This shows the power of rhetoric in science funding. The second factor is providing the right mix of *convention and originality* – a study should link to the past and uphold accepted methods, while doing something new. If a proposal is either too novel or too conservative, funding is unlikely.[4]

After a research study has been completed, there is more persuasion to be done. After submitting the report of the study to a journal, it will then be sent by the journal editor to anonymous reviewers who will be asked for their feedback and their view on whether it should be rejected, changed or accepted. In stark contrast to the dry technical language of the scientific journal article itself, peer review documents often use strong evaluative and emotive language.[5] In a positive review, reviewers may use words indicating the article to be "exciting," or "elegant"; if negative, reviews sometimes stoop to words such as "nonsense" or "useless." Frequently reviewers do not agree in their appraisals of a manuscript, which creates a rhetorical challenge for the author who must accommodate different criticisms while seeking some kind of consensus.[6]

Authors and reviewers are usually kept blind to one another's identity during the process of peer review. This standard of blind review acknowledges that if identities are known, reviewers can be biased, and that authors may be offended by the strong language that reviews contain.

Once a manuscript has been accepted following the peer review process, it enters the official corpus of science. Because most reports will contain mathematics and phrases that the average non-scientist will not understand, most people must accept the written documents of science unquestioningly, or not at all. Even for those researchers who have the knowledge and kit to test a reported scientific finding, the vast majority of their scientific knowledge will be taken secondhand from the words contained in articles and books. Given all of this, science depends enormously on the lucidity and transparency of the language used in reports to ensure that subject matter is not hidden or distorted. It also depends on the written depiction of the world in mathematical terms, and it is to this that we turn next.

Exercise for verbal precision: Writing down your own worldview

This exercise aims to use the benefits of writing ideas down to clarify your own worldview. A worldview is your set of ideas and beliefs that together define your personal way of understanding life and reality. If your worldview is in line with an existing religion or with scientific atheism, it may be easy to articulate. For the many of you who will have developed a worldview that draws on various sources and has an element of individuality, it is both challenging and rewarding to write it down. To start the process, write down your answers to the following questions:

1. What do you value most in life, and why?
2. What makes for a life well lived?
3. On what basis should you (and people in general) judge right and wrong?
4. Is there anything you would go to war for? What? If not, why not?
5. What does freedom mean to you?
6. What are the best ways of discerning fact from fiction, or truth from falsity?
7. Why is there anything at all?

An agnostic position of 'don't know' is an eminently appropriate answer to some of these questions. However, if that is your answer, try to write down why you think you don't know, and what you think is knowable and expressible about the matter. Try and write at least several paragraphs in response to each question. Be aware that this is not an easy exercise, and you may have to jot down notes before you can come up with coherent answers.

The second part of the exercise mimics the peer review process of scientific publishing. Give the document to someone else who you respect in their capacity to think about big issues, and get them to ask questions about things that they think are ambiguous or unclear, and then develop the document to improve its clarity. Then, if you feel up to it, give the document to someone who you think will disagree with your worldview in part or whole, and then critically discuss it with them. If you can continue to feel conviction in it despite criticism, then that is a very positive sign that it is coherent. You will find that your worldview develops continually as you pass through life and learn new things. That is a good sign. A healthy worldview should evolve in tandem with your own experiences and maturity.

Science and the language of mathematics

Mathematics is so fundamental and so self-justifying that it is easy to forget that it is a language. In his book *Number: The Language of Science*, Tobias Dantzig discusses how science is dependent on mathematics to function, but that this dependence does not run the other way; mathematics is not dependent on science, hence predates it historically. The symbols and rules for representing quantitative ideas in science have been drawn from various cultures and traditions over the millennia, just as the symbols of other languages have. To understand where these come from helps to bring the linguistic nature of mathematics into perspective. Our number symbols (1, 2, 3 etc.) are based on Arabic and Hindu symbols; our calculation symbols (-, +, = etc.) developed in medieval Europe; our geometrical figures

(□,○) came from ancient Greece, and the superscript numbers that we use to indicate squares of a number (e.g. 5^2) and other exponents were developed by Rene Descartes, based on the work of Pythagoras.

This culturally hybrid set of symbols provides the precise quantitative language of science, but it is far from perfect. For example, we use the base-10 system to count and measure (grouping numbers into sets of 10). Dantzig argues that this is because we have 10 fingers, so that in the days before calculators and written language, base-10 allowed people to count in sets of 10 using their fingers.[7] Other mathematicians argue that base-10 reflects a transcendent mathematical archetype called the "decagon."[8] Either way, it is just one way of grouping quantities into sets. Other bases such as base-7 and base-12 are actually better for some scientific calculations, and of course, we still use base-12 for the hours of the day.[9]

From the time of the Roman Empire up to the early middle ages, the dominant mathematical language in Europe was the system of Roman numerals. It used letters from the alphabet to represent numbers (I, V, X, L, C, D and M). It was a particularly poor system for calculation for it had no zero, and this held back mathematics for centuries. Around 1400, a brand new mathematical language emerged from the East, which would pave the way for science by facilitating new processes of calculation that were impossible with Roman numerals. The new system had been devised by the Hindus in India, and then had been adopted by the Arabs. It employed just ten symbols for 1 up to 9, and 0 for zero, and combinations of these for larger numbers.[10] The new system used the position of a symbol to convey its meaning. For example, the 1 in numbers 1, 10, 100 and 1000 mean different things. This Hindu-Arabic number language brought a whole new range of ways of doing sums and calculations using equations. Crucially, it also provided a way of doing calculus (the mathematical study of change), which in turn facilitated Newton's breakthroughs in understanding the mechanics of movement.

The idea of a square number is an example of an old idea for representing numbers that has had an important influence on modern science. It was ancient Greek mathematician and philosopher Pythagoras (570–495 BC) who first referred to square numbers. He found that some numbers when arranged as dots make the form of a square, as shown in Figure 8.1. He found these to have special properties, for example, the square of a number n is equal to the sum of the first n odd numbers (e.g. $4^2 = 16$, and $1 + 3 + 5 + 7 = 16$).

Scientists and mathematicians still refer to such numbers as squares, but use a system for representing them developed by Rene Descartes. He suggested depicting square numbers as the side length of the square with a superscript 2. This has been the convention ever since. So the numbers in Figure 8.1 can be written as 2^2, 3^2, and 4^2.

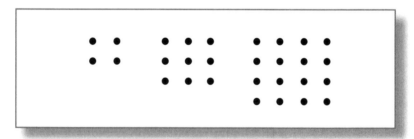

Figure 8.1 Square numbers as per Pythagoras

Square numbers are central to science, for many mathematical laws in physics contain them. For example, Newton's law of gravitation contains a square number. It states that a particle attracts another particle with a force that is proportional to the product of their masses and inversely proportional to the *square* of the distance between their centers. There is also a square in one of the most famous equations of all time: Einstein's $E=mc^2$

(which stands for the fact that the quantity of energy in a thing is equal to its mass multiplied by the speed of light *squared*). Pythagoras would have said that the prevalence of square numbers in science is evidence of a divine calculator. Galileo, skeptical of dogmatic religion as he was, still had no doubt that the reason behind the mathematical nature of science's laws was a divine intelligence who had created the universe using a pure language of quantity. He wrote of this as follows:

> Philosophy [i.e. physics] is written in this grand book the universe, which stands continually open to our gaze. But the book cannot be understood unless one first learns to comprehend the language and to read the alphabet of which it is composed. It is written in the language of mathematics, and its characters are triangles, circles, and other geometric figures without which it is humanly impossible to understand a single word of it; without these, one wanders about in a dark labyrinth.[11]

The language of mathematics is still evolving. Over recent centuries, new concepts and symbols have been added to mathematics including *imaginary numbers* and *complex numbers*, which cannot be conceptualized in the way that normal integer numbers can. Imaginary numbers are square roots of negative numbers. These have no numerical value, at least not one that can be written down. Yet if one calculates an equation that includes an imaginary number represented as an algebra term, the equation comes up with a solution, so on some level, they exist. They have practical value, for example in calculating sine or cosine waves, including wireless technologies, radar, and brain waves. Complex numbers are even stranger – they are combinations of real numbers and imaginary numbers, expressed in the form $a + bi$, where a and b are real numbers and i is an imaginary number. These also have scientific uses, for example, the wave function of a photon can be depicted as a complex number. With these esoteric concepts and symbols, mathematics has taken science right to the edge of what is possible with language and is teetering on the edge of the ineffable.

At the science-spirituality interface: Sacred geometry

Geometry is a topic that is of deep interest to both science and spirituality, and hence provides a link between the two. It is a fundamentally visual aspect of mathematics, being concerned with the figures and forms created by lines, points and the angles in two and three dimensions. From a scientific perspective, the language of geometry is used in a host of different scientific fields, including Newtonian mechanics, the refraction and reflection of light, particle accelerators that analyze the trajectories of colliding particles, medical technologies such as CT scans and MRI that use geometry to piece together images from data, and in computer models that create three-dimensional representations of natural systems. From this perspective, it is seen as a quantitative tool, no more spiritual than arithmetic.

From a spiritual perspective, geometry has been a universal inspiration for art, contemplation, and symbols in religious traditions, and now in nonreligious spirituality too. In Christianity, forms of geometry such as the golden ratio (1:1.618) have been used to design cathedrals and their windows, while in Islam, such is the reverence of geometry that almost all sacred art is based on interlocking geometric forms and mosaics. In Buddhism and Hinduism, mandalas and yantras are also widely used for spiritual contemplation and symbolism, and in esoteric Judaism, such as Kabbalah, mystical truths are conveyed by way of geometry, including its central mystical symbol called the 'tree of life.'

Geometry opens its sacred secrets when experienced intuitively rather than rationally. Seen this way, it represents eternal perfection. For Pythagoras, it was as close as humans could get to the divine nature. While scientific theories have come and gone across the centuries, geometric proofs such as the Pythagorean Theorem for calculating the hypotenuse of a triangle remain as perfectly true now as they were two thousand years ago.

Geometry also has an important curiosity in its nature as a language. The symbols used to represent geometric forms are drawn circles, squares, lines and points. These are visual analogs of the real thing, a bit like hieroglyphics. They thus directly represent what they refer to in a way that words and numbers do not. For example, a drawn circle looks like a circle, but the word cat looks nothing like a cat, and the number 5 looks nothing like what it means. In fact, one can say that a drawn circle it not just a representation of a circle, it *is* a circle. Thus do the universal and the particular merge together in the geometric shape. This is why the plaque attached to the Voyager space probe (put on the probe in the unlikely case that it was found by aliens) had its messages written in geometry, for the assumption was that geometry would hold perfectly true for nonhuman species, even though they wouldn't understand a jot of anything else we said or wrote.

Sacred geometry is the practice of working with geometrical forms to explore their deeper meanings, and to use them in artwork or architecture. Exploring the archetypal *meaning* of geometrical forms and meditating on their symmetry is said to transform consciousness and to facilitate an understanding of one's own true nature.[12]

An important symbol in sacred geometry is the *Vesica Piscis* (see Figure 8.2). It comprises two circles, drawn so that the second circle's center is on the rim of the first circle. The Pythagoreans considered it a holy figure, and female. It symbolizes the emergence of duality from unity, and the maternal power to create. By using just the lines of the form, plus the two central points of each circle, with just a compass, a straight edge and a pen, one can draw an equilateral triangle, a square, a pentagon and a hexagon, without any measuring being done.[13] The central almond-shaped section taken on its own is referred to as the mandorla. Arches in Gothic architecture are all based on this form, and it is used as the basis for the popular Jesus fish image.

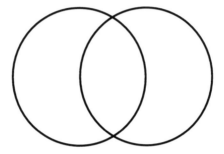

Figure 8.2 Vesica Piscis: A root form of sacred geometry

By contemplating two-dimensional geometrical forms such as the vesica piscis, or three-dimensional shapes such as the Platonic solids (the tetrahedron, cube, octahedron, dodecahedron and icosahedron), the spiritual adept is said to glean insights into the ordering nature of consciousness itself, as the ultimate ground of all such forms. *Any* form, geometric or otherwise, seen as a unified whole, is a product of both observer and observed. Take the vesica piscis – there is nothing in the image that says it should be viewed as one whole, or as two circles. That is entirely up to you; you can mentally flip-flop between the two interpretations of the lines, if you so desire.

Exercise for sacred geometry: Drawing a 'seed of life'

The seed of life, shown in Figure 8.3, is a simple but classic sacred geometry form that represents emergent form from a single geometrical shape. To draw one, you will need a geometry compass, a pencil and a piece of paper. First draw the central circle, with its center point in the middle of the page. Without changing the size of the compass opening, place the point of the compass on the rim of the circle, and make another. You will then have a *vesica piscis*.

Figure 8.3 The Seed of Life

Then use the compass to make a third circle that has as its center one of the two points where the circles overlap. Then use the same strategy for the third circle, and for the fourth, fifth, sixth and seventh, moving around the central circle as you go until you have six circles around the central one and a flower symbol will appear at the center. Then fix the compass point at the center, and the pencil on the outer edge of one of the outer circles, and draw a circle around the whole image. I recommend you then color in the central floral element, and spend some minutes gazing at it before shutting your eyes and holding the image in your mind as you meditate. The perfect harmonies of the form help to create a parallel sense of inner balance.

The East-West roots of spiritual ineffability

Wittgenstein suggested that spirituality starts where language ends. In support of this, the spiritual landscape of the twenty-first century is structured to a large degree around silence, with meditation practices, yoga, silent retreats, singing, chanting and dance practices that all involve a moving beyond language. All these lead the seeker toward inward quietness and spaciousness.[14] Silence is, from a spiritual perspective, not an emptiness but a fullness. To experience an inner silence brings a nonverbal realization that beyond words and concepts there is an indivisible ground or oneness to existence, which we experience as peace and love.[15] This path of ineffability has roots in both Western and Eastern mystical traditions, and is in many ways a synthesis of the two.

In the West, the idea that religious truth is beyond language has been central to the apophatic tradition of Christianity. The word *apophatic* comes from the Greek words *apo* (beyond) and *phania* (speech). Apophatic Christianity can be traced back to a lineage of ascetic hermits living in the Egyptian desert in the third century AD, called the Desert Fathers. They chose to locate their monastic communities in the desert, for the reason that the barrenness and silence of the landscape acted as a living metaphor for how God and Spirit are found in silent contemplation.[16]

In contrast to the complex word-based theology that has become central to mainstream Christianity since then, the Desert Fathers believed that God transcends all attributes, being beyond opposites such as good-bad or masculine-feminine, and indeed beyond all words. Following this tradition, the Roman Catholic nun and Nobel Peace Prize winner Mother Teresa wrote that God is found in silence, which is in turn found in contemplating nature:

We need to find God, and he cannot be found in noise and restlessness. God is the friend of silence. See how nature –

trees, flowers, grass – grows in silence; see the stars, the moon, and the sun, how they move in silence. We need silence to be able to teach souls.[17]

In recognition of the limits of language, the spiritual practices of the Desert Fathers were characterized by silent contemplation. It was said that one, Abba Agathon, kept a pebble in his mouth for years to aid his pursuit of spiritual growth through silence. Another story from this tradition is that when a high-ranking Christian priest came to the monastery of Abbo Pambo for guidance, the monk remained completely silent in order to teach him that holiness cannot be found in words.[18]

The apophatic tradition of the Desert Fathers experienced a brief revival in medieval Europe, thanks in part to the works of Meister Eckhart (1260–1328). Eckhart was a philosopher and mystic who gave new life to ancient apophatic ideas. He wrote again and again that spiritual truth transcends language, for example, that: "God, who has no name – who is beyond names – is inexpressible."[19] Written briefly after Eckhart, *The Cloud of Unknowing* continued this revival of the apophatic tradition, emphasizing the limits of intellect and language in approaching God. It argued that the gulf between God and human language is so great that we cannot say that God is good, or male, or even that he exists. All of these have opposites, and the apophatic God is beyond opposites.[20]

From the seventeenth century, the apophatic Christian tradition was given a new expression by the Quakers, who from the outset were skeptical of words and theology. Instead, they prioritized an experiential quest to understand the inner relationship between God and the soul, over and above beliefs and scripture. This apophatic ethos that words are fundamentally limited is still apparent in Quaker writings. In a 2004 publication *Twelve Quakers and God*, it is stated that "… to define or describe God is to distort, to impose our own limitations of time and space,"[21] and "from very young I have been awe-struck by experiences I had no name for. As I grew up I came to understand these in terms of God."[22] Reflecting this belief that words are always provisional in spiritual matters, the key Quaker book *Faith and Practice* is updated and added to in each generation, so allowing their verbalized belief system to evolve over time, rather than becoming a fixed scripture.

The Sufi mystical tradition has its roots in the desert landscape, and like the Christian Desert Father tradition, it emphasizes the power of silence. Rumi, the prolific 12th-century Sufi poet, wrote of the importance of moving past spoken or written language into silence to cultivate spiritual knowing, and thus being ready to let go of theology:

> You and I have spoken all these words,
> But for the way we have to go,
> words are no preparation,
> I have one small drop of knowing in my soul.
> Let it dissolve in your ocean.
> This love
> Is beyond the study of theology
> That old trickery and hypocrisy.
> If you want to improve your mind
> that way, sleep on.[23]

In the nineteenth century, the ideas of Buddhism, Taoism and Hinduism entered into Western consciousness for the first time, as teachers and texts poured in from the East. They provided an important new impetus to the search for ineffable spiritual truth beyond words, for all three of these traditions emphasize the limits of language. The *Tao Te Ching* states that language cannot reach the Tao, which is the ultimate essence of all. It states: "The Tao that can be told is not the eternal Tao; the name that can be named is not the eternal name." A similar view is found in Buddhist scripture. The Chinese Buddhist spiritual text *The Secret of the Golden Flower* states: "Only through meditation and quietness does true intuition arise... All methods end in quietness. This marvelous magic cannot be described."[24] Similarly, the *Lankavatata Sutra* states: "When appearances and names are put away and all discrimination ceases, that which remains is the true and essential nature of things and, as nothing can be predicated as to the nature of essence, it is called the *Suchness of Reality.*"

There is an old adage in Buddhism that words are like a finger pointing to the moon, not the moon itself. To really *see* the moon, in a way that leaves words and their preconceived meanings behind, is the ineffable path of spirituality. It is to encounter reality as it was before language and will be after language, in all its primal originality.

In the Rig Veda, a Hindu scripture going back to 1000 BCE, God is described as *Asat*, which means silence. In the Mandukya Upanishad, silence is equated with the soul itself, thus to know silence is to know the soul, which is in turn to know God. Showing parallels with the aforementioned story from the Christian monk Abbo Pambo, the eighth century Hindu philosopher Shankara wrote how a great master, when questioned about Brahman (God or Ultimate Reality) by a student, became silent. When the student repeated his question three times, the master replied, "I am teaching you indeed, but you do not understand. Silence *is* the answer." Following this lineage, the twentieth-century Hindu sage Rabindranath Tagore wrote: "The water in a vessel is sparkling; the water in the sea is dark. The small truth has words that are clear; the great truth has great silence."[25]

Some meditative techniques aim to bring about inner silence by the use of a mantra – a word or phrase that is repeated inwardly or out loud during meditation. Mantras paradoxically use language to transcend language, for, in the repetition of a single word or phrase for a period of minutes or hours, the sound loses its associations and meanings. This kind of practice is perhaps most strongly associated in contemporary spirituality with Transcendental Meditation, which employs the repetition of Sanskrit mantras to create a state of inner silence. The West also has a rich heritage of mantra meditations through its apophatic tradition. The Desert Fathers of Christianity developed a practice termed *hesychasm* that involved repeating a simple phrase inwardly, while in outward silence, for many hours on end. Similarly, *The Cloud of Unknowing* proposed intensive contemplation by repeating a single word such as *God* or *Love* for hours on end.

The modern mystic Jiddu Krishnamurti, who was highly influenced by Buddhism, writes of the silence brought about through meditation as one that can't be constructed in thought: "A meditative mind is silent. It is not the silence which thought can conceive of; it is not the silence of a still evening; it is the silence when thought – with all its images, its words and perceptions – has entirely ceased."[26]

Modern spirituality has inherited the aim of the mystic to move beyond words into a deeper experience of self and world. That is not to say that there is a lack of literature on spirituality, for there is much. It is to say that all such words are signposts beyond themselves, beckoning the practitioner towards truths that are beyond words and belief. The cultural consequence of this pursuit of the ineffable over the centuries has been a prolific outpouring of music, art and poetry that all endeavor to somehow represent or point towards that which lies beyond language.

Exercise for silent awareness: A silent day

Put a day aside to see what it is like to not talk or read for a day. Don't stay inside – walk around somewhere and observe people and the world in silence. You will likely notice that the amount of verbalized thought in your mind decreases over the course of the day, and with it, worries may lessen in intensity. You will become aware of subtleties in your environment as your mind attunes more to your senses instead of abstract thoughts, and you may start to think about other people more. If you feel moved to continue this practice, consider signing up for a silent retreat.

Music and art as expressions of the ineffable

The modern experiment in expressing the ineffable through artistic or musical means is one of spirituality's greatest gifts to society. Spiritual seekers across the past centuries, in attempting to express that which the formal language of science and philosophy cannot, have given us some of the great art and music of the era.

In his book *Music and the Ineffable*, Vladimir Jankélévitch argues that music has a unique power to draw the mind towards ineffable truths beyond words, for it conveys a deep meaning and feeling in the listener without necessarily using any words or representing anything directly.[27] Composers during the modern era have been released from religion and theology to convey music's intrinsic message of meaning beyond language. This has become a cultural mainstay of the Western world, with people of all ages listening to secular music of all genres to experience an unmediated sense of self-transcendence and beauty. The Romantic composers sought inspiration directly from the world's mystical traditions to achieve this feat. Ludwig van Beethoven, for example, was influenced by both Hindu and Buddhist mysticism and their emphasis on experiencing and expressing a truth beyond language.[28] His later works particularly were composed to express the ineffable unity that lies behind the expressible diversity of phenomena.[29]

Ralph Vaughan Williams (1872–1958), composer of works including *The Lark Ascending* and *Fantasia on a Theme by Thomas Tallis*, described his music as a contemplative exploration of spirituality beyond language. He has been described by a biographer as a "spiritual vagabond" in his tendency towards spiritual exploration and his preference not to commit to one religion.[30] In leaving behind the systems and theology of religious doctrine, he aimed for his work to convey a revelation of the sacred that is beyond labels and the divisions of belief.

As well as composers, famous musicians of the past century have been central to music's deployment as a spiritual tonic for overcoming language and its constraints. The jazz musician John Coltrane followed no creed or religion, but was deeply spiritual and sought to convey the divine in his music beyond language. He said that "music can make the world better and, if I'm qualified, I want to do it. I'd like to point out to people the divine in a musical language that transcends sounds. I want to speak to their souls."[31]

Within contemporary music, the genre of electronic ambient music can be seen as an expression of the continued search for silence and the ineffable that follows in the footsteps of the aforementioned composers, while taking a more explicitly meditative approach. Following Vaughan Williams, ambient music avoids lyrics in order to allow the mind into more expansive spaces than those structured around linguistic thought. The genre has developed alongside the growing popularity of meditation. For example, Mike Oldfield, the composer of *Tubular Bells*, is a prime example of the interaction between meditation and ambient music. He describes his experiences in meditation as entering pure "empty space" that is beyond opposites, and how this has inspired him to create music that takes the listener into spaces of consciousness beyond the ordinary.[32] Spiritual practice and electronic ambient music have been further intertwined over recent decades by the development of music that is explicitly designed to support meditation, such as the work of Christopher Lloyd Clarke. I myself regularly use his one-hour-long tracks such as *Adrift, Ascension* and *Untold Depths* as an audio accompaniment to meditation.

Alongside music, the visual arts have been explored as ways of expressing the ineffable during the modern era, starting with the art of the Romantics, then the Impressionists and Symbolists, and then abstract modern art. Kandinsky and Mondrian, early pioneers of abstract art, were both committed to the spiritual movement of Theosophy and to the practice of meditation that is central to it. Mondrian wrote that "art gives visual expression to the evolution of life, the evolution of spirit and – in the reverse direction – that of matter."[33] The abstract images he and Kandinsky produced were stimulated by a search for truths that reached deeper than language. They intentionally rendered the viewer speechless, either through admiration, wonder or puzzlement. In his book *The Search for Meaning in Modern Art*, Alfred Neumeyer wrote on the importance of ineffability in Kandinsky's work:

> ... because Kandinsky and most of the painters after him do not refer to reality but instead produce a reality which has not existed before, our language actually lacks a vocabulary with which to describe such painting... The meaning of modern art may be formulated, but its expression cannot be translated into the medium of words... Whatever Kandinsky may have expressed through his dots, lines and colors, we participate in an awakening of sensuous-spiritual phenomena.[34]

Modern abstract art thus emerged within a spiritual, but not religious, context, and so can be seen as one of the many important manifestations of the modern search for expressing that which is beyond the rational and empirical.

Ineffability and religious tolerance: Meeting in the oneness beyond language

Ineffable spirituality is not just powerful fuel for inspiring music and art; it also contains a message for how we can relate to each other in a plural world that contains different religious and nonreligious worldviews. The religious conflicts that have been a defining feature of the past thousand years are evidence of what happens when verbalized systems of belief are erected around ineffable truths, and then taken to be indisputable fact rather than signposts pointing the way beyond themselves.

Scriptures and theologies across the religions are different in many ways, and if one were to adjudicate between religions on the assumption that one such verbal formulation must be wholly true, the others would have to be rejected. By taking such a viewpoint, the religious zealot or fundamentalist sees a world divided between the saved (those who accept the single true scripture and the tradition built on it), and the damned (everyone else).[35] Such a view is both destructive and wrong. It is destructive because it leads to fear of the other, and arguably justifies any actions to bring others around to one's own religion, for anything is better than hell. It is wrong for many reasons, not least because words are signs that sit within endless layers of interpretation. Words are dependent on the meanings of the person who produces them, the interpretations of the person who receives them, and on the cultures and philosophies from which producer and listener draw their understandings.

Extra layers of interpretation and ambiguity are created when text is translated between languages and over historical epochs, as all scriptures are. To take just one example, the New

Testament verse Luke 17:21 is translated from the Ancient Greek in the King James Bible as "Neither shall they say, Lo here! or, lo there! for, behold, the kingdom of God is within you." In the International Standard Version the passage is translated "People won't be saying, 'Look! Here it is!' or 'There it is!' because now the kingdom of God is among you." In the King James version, the translated meaning points to the mystical inner path to the divine, in the other, it does not at all. Hence can single translated words flip the entire meaning of a scriptural passage.

A breakthrough in realizing that the unifying core of religion lies beyond words occurred in the late nineteenth century at the 1893 Parliament of the World's Religions in Chicago. The assembled gathering, one of the first ever to bring luminaries of different faiths together, agreed that the mystical paths within different religions have remarkable similarities. It became increasingly apparent from that time onwards that the spiritual *experiences* across different religions have likenesses, even though the word-based theologies differ. Many classic works have been written exploring this thesis, including Aldous Huxley's *The Perennial Philosophy*, and more recently Wayne Teasdale's *The Mystic Heart*.[36,37] The common denominators that this inter-faith approach highlights within all mystical and inwardly reflective forms of religious experience are: the ultimate value of love; the experience of oneness; gratitude and reverence; a deep sense of communion with others (human and nonhuman); an ethic of compassion and care; practices for clarifying consciousness; and the conviction that spiritual truth lies beyond the reach of words.

Chapter 9

Explanation – Contemplation

In 1901, the American philosopher William James was invited to Edinburgh to deliver the Gifford Lectures. This lecture series had been developed as a platform for natural theology, which promotes reason and argument as the way to God. So it was something of a shock to the Scottish audience when James stated that his topic for the lectures was mysticism. The meditative and ecstatic practices that mystics use to seek the divine are in many ways the polar opposite of the spiritual approach to the logical arguments of natural theology. Despite this contrast with the Gifford Lectures' normally sober material, the lectures were so fascinating that the Scots were won over. During the lectures, and the book *The Varieties of Religious Experience* which was based on them, James at various points considers the complementary relationship of science and spirituality (which he refers to as "personal religion"). A key reason for their complementarity, according to James, is that science seeks to *explain* reality via general concepts, law and principles, while spirituality deals primarily with the kind of knowing that comes through *contemplation* of the raw, felt experience of life in its direct particulars.

James argues that these two kinds of knowing – the explanatory and the contemplative – support each other despite being juxtaposed.[1] Explanation seeks to find the cause or reason for a phenomenon, by looking into the past for possible causes, or into the abstractions of mathematics and reason for a law or principle that might explain or govern it. The Latin root of the word explanation is *ex planus*, which means 'to spread out', and this reflects how in the explanatory mindset attention spreads outward and away from a phenomenon, to seek out causes, laws and reasons for why it is as it is.

In contrast, contemplation involves bringing sustained attention to a particular object or image and immersing oneself fully in the experience of it. This requires being fully in the present moment, and placing attention as directly as possible on that which is contemplated. This, in turn, means *not* trying to work it out or explain it. The Latin root of contemplation is *com templum*, which means 'together in the sacred place.' This reflects how the perceiver and the perceived come together in the act of contemplation in a higher sacred unity.

Following James' lead, this chapter explores how science is characterized by explanation and abstraction, while spirituality is characterized by contemplation and practices that involve deep immersion in particulars. At the interface between these, I look at three immersive methods within the social sciences – phenomenology, co-operative inquiry and ethnography, as examples of where the explanatory ethos of science and the contemplative ethos of spirituality have merged.

Science and the pursuit of explanation

All branches of science aspire to create general theories that explain phenomena as manifestations of a fundamental law, principle or pattern.[2] Theories can be descriptive or explanatory. Descriptive theories tend to be called models and are assumed to be a first step on the path towards the goal of explanation. Explanatory theories propose a general principle or law that explains *why* (and how) a natural phenomenon happens, and hence a scientific explanation of a thing involves subsuming it under such a general principle or law.[3] Some theories achieve both goals of describing and explaining phenomena within a unified scheme, and one of the most successful examples of this is atomic theory.

The Nobel Prize-winning physicist Richard Feynman argued that atomic theory is perhaps the most powerful unifying hypotheses in the history of science.[4] Formulated initially by John Dalton in the 1800s, and then developed by Ernest Rutherford in the early twentieth century, atomic theory states that all things in the universe are made of atoms, which move around constantly and are composed of electrically charged parts. All atoms contain a nucleus, in which there are neutrons (neutral charge) and protons (positively charged). Electrons (negatively charged) are in orbit around the nucleus. Four fundamental forces act on atoms – gravity, electromagnetism, the strong interaction (which holds nuclei together), and the weak interaction (responsible for radioactive decay). There are different kinds of atoms, called elements, each of which has a different number of protons in the nucleus, and these are all listed in the periodic table. Elements combine into molecules, and chemical reactions occur when atomic arrangements in molecules change.

These basic tenets of atomic theory can be used to describe and explain the behavior of an enormous variety of phenomena in non-living and living things, including the following:

- Heat is explained as the motion of atoms; if energy is applied to a substance, increased atomic motion leads to more collisions between atoms, and this brings about a rise in temperature.
- Pressure is explained as the density of atoms in a space, with pressure increasing as more atoms crowd into the same three-dimensional space.
- States of matter – gases, liquids and solids – are explained as different spatial configurations of atoms, with density of any element being lowest in its gas form (atoms furthest apart), more dense in liquid form (atoms closer together) and most dense in solid form (atoms closest together, apart from water, whose liquid form is denser than its solid form, hence ice floats).
- Molecular bonding is explained as atoms arranging to

find an electrically neutral structure, where the pluses and minuses of the component parts balance out. Ions are molecules that do not electrically balance out.

- It explains why chemical reactions proceed as they do, on the basis of how electrons are transferred between atoms.
- In living things, all of the physical senses are understood as the interaction of moving atoms, with the exception of vision which depends on photons. Smells can be explained as the interaction of particular molecular combinations of atoms with our noses, and the same is true of taste on the tongue. Touch is the product of pressure between the skin and other material surfaces, the atoms of which repel each other when pushed together. Hearing is the result of oscillating motion of the atoms of the gases in the air colliding with the eardrum.
- In molecular biology, all theories rest on atomic theory. For example, the auxin theory of how flowering plants move to follow the sun, developed in the 1920s by Nikolai Cholodny and Frits Warmolt Went, states that a class of growth hormones called auxins are more active in the shade than in the sun, so the shady side of a plant grows more than the sunny side, leading to growth that tips the stem and leaves towards the sun. Auxins are molecules composed of the elements of oxygen, hydrogen and nitrogen, which in turn are comprised of atoms in combination. When plants move towards the sun, the atoms rearrange into a new form.[5]

The philosopher Carl Hempel referred to the process of explaining particular things by recourse to a general theory like atomic theory as *deductive-nomological explanation*.[6] For example, an explanation to the question "why did the apple fall when dropped?" according to atomic theory and Newton's law of gravitation would be as follows: Objects composed of atoms attract each other due to gravitation, with a force proportional to their mass and inversely proportional to the square of their distance; the apple and Earth are such objects, they are therefore attracted to each other; given that the Earth is so much more massive than the apple, the apple falls towards it when the inertia of being held is removed, and the Earth's movement towards the apple is negligible.

Exercise for practicing scientific explanation: Explaining the outcome of a home chemistry experiment using concepts from atomic theory

The process of scientific explanation involves invoking theories and laws to make sense of particular events. A simple experiment you can do to practice scientific explanation based on atomic theory is the vacuum candle experiment. This activity works best if you have a small group of people present who can offer different explanations for what happens, but you can do it on your own too.

Put a candle upright in the middle of a large bowl (stick it down using some melted wax from the candle), then add in about 5 mm depth of any colored liquid to the bowl. Then put a glass or glass bottle over the candle and then watch what happens to the candle and the level of the liquid in the glass.

After observing the outcome, try to explain it by your understanding of atoms, temperature, pressure and states of matter. You will need to explain it by means of general principles, but you and your friends may not agree which one(s). The actual explanation based on atomic theory (which is more complex than you might think) is given in the endnote provided here.[7]

Hundreds of other phenomena across the sciences could be listed that are also founded on the axioms of atomic theory, yet the theory is conjectural, for no one has ever seen the structure of an atom. A technique called scanning tunneling microscopy can show atoms as hazy blobs, but nothing more. They are just too small. Indeed, atomic theory may be more of a powerful metaphor than a statement of fact, for quantum physics now shows that the tiny particles that comprise atoms behave like waves, can be smudged out in space and time as parts of a field, and don't seem to have a fixed location until measured. So the theory of atoms is really a general scheme that humans have imputed, with great success, to make sense of reality in ways that we can apply. To explain a thing by way of atomic theory or any other general scheme is a very human process – a way of thinking – that has roots going all the way back to Plato.

Why the general is assumed to explain the particular

A common phrase for making sense of why scientific laws explain particular phenomena is that the former *govern* the latter. To give one example amongst many of the use of this phrase, in the below passage Richard Feynman writes of how the law of the conservation of energy governs all natural phenomena:

> There is a fact, or if you wish, a *law*, governing all natural phenomena that are known to date. There is no known exception to this law; it is exact, as far as we know. The law is called the conservation of energy; it states that there is a certain quantity, which we call energy, that does not change in manifold changes which nature undergoes. That is a most abstract idea, because it is a mathematical principle; it says that there is a numerical quantity, which does not change when something happens.[8]

This idea that abstract laws govern particulars goes all the way back to Plato's theory of forms, and to Francis Bacon's reinterpretation of this theory for modern science. Plato said that reality is comprised of both a physical level of particulars *and* a higher archetypal level of general forms. Forms are ideals of Truth, Goodness and Beauty that exist out of time and space in a transcendent mind. Importantly, forms are more real than particular objects, and have a causal power over them. They influence the physical level of reality and can be said to *govern* over it. A scientist who is unusually explicit about science's underpinnings in Plato and Bacon is the physicist Roger Penrose. In his 2005 book *The Road to Reality: A Complete Guide to the Laws of the Universe*, Penrose writes that he is open to the idea that reality may have a Platonic level of form that includes mathematical laws. He writes:

> Platonic existence, as I see it, refers to the existence of an objective external standard that is not dependent upon our individual opinions nor upon our particular culture. Such 'existence' could also refer to things other than mathematics, such as to morality or aesthetics, but I am here concerned just with mathematical objectivity, which seems to be a much clearer issue... Plato himself would have insisted that there are two other fundamental absolute ideals, namely that of the Beautiful and that of the Good. I am not at all adverse to admitting the existence of such ideals, and to allowing the Platonic world to be extended so as to contain absolutes of this nature.[9]

Plato illustrated how the form level of reality is more real than the physical level in his allegory of the cave. In the allegory, presented in his work *The Republic*, prisoners are imagined who have been in a cave for their whole lives, and who can only see a wall in front of them. On this wall, shadows appear from figures outside the cave. For the prisoners, the shadows are all they know of reality. However, the shadows are in fact just images cast by the real figures outside of the cave. So it is with human life, said Plato, we are like prisoners in the cave of our senses and thoughts. What we see are shadows of the real thing that we can never see directly, and these real things are not concrete objects but perfect ideas. Like the prisoners raised in the dark of the cave, if we were taken into the light of the higher realm of form and archetype, we would be blinded by it.

Francis Bacon reinterpreted Platonic forms in order that mathematical laws could be considered examples of such forms. Plato did not make this assumption, but it fitted well with Bacon's mechanistic agenda for science.[10] Bacon suggested that scientific laws, framed quantitatively, actually *govern* particulars in a semi-divine way, a bit like God governs his creation. They are typically formulated as equations, such as Newton's second law of motion *f=ma (force = mass x acceleration)*, Hooke's law of spring extension *F=kX (force = spring stiffness x distance spring is stretched)*, and Einstein's $E=mc^2$ *(energy = mass x the speed of light squared)*. Within such laws are physical *constants* – numerically defined parameters that are set at a fixed value. There are 26 constants in the Standard Model of physics, including the speed of light, the Newtonian constant of gravity, the electric constant, and the cosmological constant (the energy density of the vacuum of space). They are necessary ingredients in many scientific laws, for example, the *c* in $E=mc^2$ is the speed of light constant (299,792,458 m/s). Nobody knows why physical constants are set at the values they are, and this means that natural laws contain the unexplainable.[11]

In summary, to do science is to be immersed in the mental rigors of explanation and in working out why and how things happen in the present as a result of hidden mechanisms, or of causes in the past. While that is an excellent recipe for general knowledge, it is not conducive to living life in a fulfilling way, for to be engaged in explanation is to have attention away from the concerns and rewards of the here-and-now. Spirituality, as science's natural counterpoint and foil, explores methods of contemplation that actively avoid explanation, and so bring attention fully into the concrete and sensuous present.

Spiritual contemplation: Immersion in particulars

Contemplative practices involve paying attention to the concrete here-and-now. Through art, music, dancing, painting, singing, chanting, drumming, meditating or yoga, a person is drawn to paying attention to the present, in all its lived particularity. Such practices have the benefit of being relaxing but can also provide deep insights too into the nature of reality in ways that explanation can't. What appears to the thinking mind as a particular object shows itself in deep contemplation to be an expression of the dynamic unity of the cosmos that transcends the subject-object split.

As with so much of today's spirituality, Romanticism set a key precedent in matters of contemplation. For the Romantics, contemplating nature and art was the very foundation of the spiritual life. Artworks and natural scenes could inspire a sense of grace, beauty or unity if perceived patiently and fully, unencumbered by explanatory concepts and judgments.[12] Being fully present *with* an object, immersed in it such that one loses a sense of separation from it, was the true way to know it. Whether a sunset, the smile of a baby, the waving of a tree in the wind or a beautiful piano sonata, the Romantics suggested that any event or object can be seen as sacred and holy through the contemplative lens. The Romantic poet William Blake provided one of the most oft-cited expressions of this contemplative insight, when he wrote:

> To see a world in a grain of sand,
> And heaven in a wild flower,
> Hold infinity in the palm of your hand,
> And eternity in an hour.[13]

Goethe wrote that contemplation could awaken a higher intuition of nature's unified order. In Goethe's approach, what appears as a particular and unique object to reason can be encountered through contemplative intuition as a seamless part of nature that contains the essence of the whole. When the philosopher Henri Bortoft tried Goethe's method of contemplating a flowering plant, it elicited an experience of the plant as a flowing oneness, rather than an object composed of parts and categorized through concepts.[14] He described awakening to this truth as seeing the "universal shining in the particular."[15] For Goethe, such contemplation should be partnered with rigorous and detached observation, so that the two can balance each other out and lead to a higher level of knowing than either can alone.

Romantics like Goethe inherited their love of contemplation partly from the traditions of the East. The flower sutra story in Buddhism describes how the Buddha stood beside a lake on Mount Gridhakuta and prepared to give a sermon to his disciples. He noticed a lotus flower in the water nearby, plucked it and showed it to all his disciples. This *was* his sermon – no words, no explanation, just a flower. The one disciple who understood the teaching is said to have founded the school of Zen, a Japanese tradition of Buddhism that places a particularly high value on meditation and intuition. The philosopher Schopenhauer, who was a key influence on the age of Romanticism, was one of the first philosophers to embrace ideas from Buddhism, and to practice their techniques of contemplative absorption. He describes here how contemplation of a particular scene or object can lead to merging of self and object into a higher form of knowing:

If, raised by the power of the mind, a man relinquishes the common way of looking at things... if he thus ceases to consider the where, the when, the why, and the whither of things, and looks simply and solely at the what; if, further, he does not allow abstract thought, the concepts of the reason, to take possession of his consciousness, but, instead of all this, gives the whole power of his mind to perception, sinks himself entirely in this, and lets his whole consciousness be filled with the quiet contemplation of the natural object actually present, whether a landscape, a tree, a mountain, a building, or whatever it may be; inasmuch as he loses himself in this object (to use a pregnant German idiom), i.e., forgets even his individuality, his will, and only continues to exist as the pure subject, the clear mirror of the object, so that it is as if the object alone were there, without anyone to perceive it, and he can no longer separate the perceiver from the perception, but both have become one, because the whole consciousness is filled and occupied with one single sensuous picture.[16]

Practices for contemplation: Sensory meditation and haiku

To truly rediscover the sacred power of the particular is to explore the various layers of mind and consciousness that lie beyond thinking. For most people, conceptual thought with its general theories and schemes is like a tap that the modern mind has forgotten how to turn off. As we have become increasingly stuck at the abstract level of thought, we have become divorced from the concrete reality of our bodies and emotions.

The contemporary mystic Eckhart Tolle proposes that immersive contemplation of the world can remedy this modern malaise, by helping consciousness out from behind its veil of thought. A meditative method that Tolle proposes involves attending deeply to a sensory experience in a non-judgmental way. This process, if pursued with discipline over time, allows the practitioner to experience states of consciousness that provide for a feeling of being more truly alive than usual, precisely because one is living in the particulars of the now, not in lifeless abstractions. Here are two extracts from his work that refer to this practice:

Use your senses fully. Be where you are. Look around. Just look, don't interpret. See the light, shapes, colors, textures. Be aware of the silent presence of each thing. Be aware of the space that allows everything to be. Listen to the sounds; don't judge them. Listen to the silence underneath the sounds. Touch something – anything – and feel and acknowledge its Being. Observe the rhythm of your breathing; feel the air flowing in and out, feel the life energy inside your body. Allow everything to be, within and without. Allow the 'isness' of all things. Move deeply into the Now. You are leaving behind the deadening world of mental abstraction.[17]

Watch an animal, a flower, a tree, and see how it rests in Being. It *is* itself. It has enormous dignity, innocence, and holiness. However, for you to see that, you need to go beyond the mental habit of naming and labeling. The moment you look beyond mental labels, you feel that ineffable dimension of nature that cannot be understood or perceived through the senses. It is a harmony, a sacredness that permeates not only the whole of nature but is also within you.[18]

Altered states engendered by psychedelics or other means of trance also draw the mind out of abstraction into the directness of the senses and the sacredness of the mundane, facilitating the kind of sensory meditation of which Tolle speaks. Aldous Huxley's work *The Doors of Perception*, written about his psychedelic experience using mescaline, describes his deep contemplative connection with a vase of flowers:

I was sitting in my study, looking intently at a small glass vase. The vase contained only three flowers... I was not looking at an unusual flower arrangement. I was seeing what Adam had seen on the morning of his creation – the miracle, moment by moment, of naked existence... what rose and iris and carnation so intensely signified was nothing more, and nothing less, than what they were – a transience that was yet eternal life, a perpetual perishing that was at the same time pure Being, a bundle of minute, unique particulars in which by some unspeakable and yet self-evident paradox, was to be seen the divine source of all existence.[19]

Another spiritual practice that also requires an open and total absorption in particular phenomena is haiku poetry, which originated in Japan. The process of haiku involves entering a still, meditative state of consciousness, and then bringing forth a brief poem that conveys a deep truth in a way that transcends reason and theory. The haiku poem is composed of three lines, the first of which has five syllables, the second of which has seven, and the last of which has five. The poem typically conveys a fleeting moment in nature. Here are two famous ones:

An old silent pond…
A frog jumps into the pond,
splash! Silence again.
(Matsuo Basho)

Over the wintry
forest, winds howl in rage
with no leaves to blow.
(Natsume Soseki)

From a scientific perspective, these poems have no merit at all as a representation of truth, for they contain no generalities, systematic description or explanation. Yet from an intuitive level of consciousness, the haiku is said to provide the receptive mind with a higher illumination beyond thought. The European poet Drago Stambuk, a devotee of haiku, writes of this:

By definition, the haiku is brief; even tiny. In three short lines, a vast metaphysical expanse opens up: this is the paradox, the shock and the achievement of haiku. A minimal amount of text charged with infinite space and time.[20]

Stambuk speaks from his experience that in haiku consciousness divisions are overcome – internal and external become one. Rather than looking at the world, the haiku writer becomes a living embodiment of the world's unity and becomes the voice of the All. Pursuing the art of haiku leads, says Stambuk, to a sense of harmony with nature and to the universal values of compassion and love that spill naturally from the heart-led intuitive mind.

While contemplative spirituality has a strong focus on the deep immersion in particulars, it is also bound in theories and explanations, albeit to a lesser degree than science. Hundreds of concepts and ideas have been developed around topics such as spiritual experiences, spiritual healing, spiritual reality and spiritual development. Reading these theories and ideas is to spirituality what reading a menu is to eating a meal. Both are useful preludes and helpful guides to the act, but non-nutritious in themselves. Ultimately, the activities of spirituality that bring healing and insight involve leaving generalities and explanations behind to move more directly into the experience of the here-and-now.

Exercise for developing contemplative awareness: Writing haiku

Creating haiku poetry is a wonderful way to allow the mind to fall out of abstraction and interpretation and into direct contact with the reality of the present moment. To start the process, spend a while in nature (this can be a garden, park, the woods or the wilds), and wait for your mind to become relaxed and open. Try to breathe in the world around you, feeling yourself to be a participating part of the environment you are within. Walk without an agenda or plan, allowing yourself to follow your intuition and curiosity. Note how you observe more than you usually do, almost as though you have taken blinkers off.

Once you feel calm and anchored in the present moment, without consciously deliberating too much, allow three nouns that describe some aspect of the scene in front of you to enter your mind. Then around these words create three short lines that have five, seven and five syllables. Don't consciously try to make the lines connect. Just focus on each one at a time. Then write the haiku down and read it back. You will be surprised at the effect. It is very hard to describe, but you may find that somehow the particularity and brevity of the haiku form creates a feeling of oneness between self and world. It is, in my experience, something that has to be experienced. As with so much that is spiritually meaningful, the experience of haiku illumination is very hard to describe in words.

At the science-spirituality interface: Immersive methods in the social sciences

The social sciences of psychology, anthropology, and sociology have brought the methods and values of science to the study of human beings. Although the natural sciences generally agree how science should be conducted, there is no such consensus in the social sciences. For over a century now there has been a range of competing methods, which vary in the extent to which they emphasize detached explanation or immersive contemplation.

The standard quantitative method of the social sciences

follows the mold of the natural sciences. It prioritizes detached observation, measurement, and clear causal explanations based on mathematical formulae and rational theories. It assumes that objective knowledge is possible that transcends the point of view of the researcher. If you were to leaf through a journal that adheres to this methodology, such as *Psychological Science*, you would not see any actual people mentioned at all. Rather than talking about individuals and their lives, reports in such journals present numerical data from a whole sample of people, averaged out and compounded into graphs and tables. In that way, they remain detached from their subject matter and aim for replicable, quantitative facts.

The merits of this approach to the study of human beings have been questioned since they were developed in the nineteenth century, and alternative methods have been developed that employ a more contemplative and immersive approach. In these methods, the researcher enters into dialogue and relationship with the people they are studying, participating in their life for a period of time until they may understand it empathically, and then uses qualitative data rather than numbers and measurement. One of the first to take this approach was William James, whose aforementioned book *The Varieties of Religious Experience* reports on a study of mystics during which he immersed himself in the practices of the mystic to gain a first-person insight into their point of view, and then studied their writings to attempt to make sense of the experience descriptively.

Since James, a number of methods have continued this approach based on deep immersion in, and personal connection with, the experiences of others. Here I will describe three of these: *phenomenology, co-operative inquiry* and *ethnography*, and shed some light on how they all link to spirituality, before discussing how the study of spirituality has benefited from both the detached quantitative approach and the immersive qualitative approach.

Phenomenology, developed originally by the philosopher Edmond Husserl, rejects objectivity and the importance of explanation, and aims instead to describe how reality and meaning are framed within the field of lived, conscious experience. *All* beings and things are connected together in consciousness, says phenomenology, and this has spiritual implications. The phenomenologist Erazim Kohak has argued that consciousness, as the fundamental "ground of being", may be what religions have traditionally called God.[21]

Phenomenology research seeks to understand lived, conscious reality by questioning all preset assumptions about a topic of study, and aims to make sense of a person or group without too much prior judgment. The researcher enters into an empathic relationship with the person or persons under study, to gain an insight into their first-person field of experience.

I personally did my doctoral thesis using a method based on phenomenology, called *Interpretative Phenomenological Analysis* (IPA).[22] I researched how and why people experience major personal and existential crises in their early adult life. During the process of interviewing participants, I could feel that the research was changing *me*, as well as acting as a data collection exercise. I found myself increasingly intertwined with the research and inseparable from the analysis and outcome; I was merging with my subject matter. From a hard-science point of view this is a source of bias, but from a phenomenological point of view it is an insight into how the self can never be isolated from its context, and that includes the researcher too.

Another immersive social science methodology that has close links with spirituality is *co-operative inquiry*. It was developed in the 1970s by John Heron and Peter Reason, in conjunction with spiritual ideas from transpersonal psychology, including the idea that reality is a participative co-creation between conscious individuals and a cosmic mind.[23] The basic premise of the method is that research should be done co-operatively *with* people as creative beings, rather than *on* people as objects. In a co-operative inquiry, a research question is explored through actions and reflections undertaken by a group of people, all of whom are recognized as experts *and* participants. The researcher thus enters into an immersive and contemplative experience that removes the clear distinction between subject and object.[24] Heron and Reason describe this as follows: "participation is a way of knowing in which knower and known are distinct but not separate, in an unfolding unitive field of being."[25]

A third immersive method that has been an important influence across the social sciences is *ethnography*. It studies culture, by way of the researcher becoming an observing *participant* within a culture for a period of time. The ethnographer is trained to see through their own background assumptions, and to gain insight into the often unspoken meanings and rituals of the culture being studied. The outcome of such research is not so much an explanation of the culture, but rather a thick description of what is like to live within it.

Ethnography has been integral to developing an understanding of how spirituality is practiced in cultures that have not adopted scriptural religions. The work of anthropologist Michael Harner is a case in point. He spent fourteen months doing ethnography with the Jivaro people of Peru and, during his research, he participated in shamanic rituals. He was impressed by the direct and powerful nature of the experiences that the rituals engendered, and felt that modern society could learn much from them. He wrote:

> One reason for the increasing interest in shamanism is that many educated, thinking people have left the Age of Faith behind them... Second-hand or third-hand anecdotes in competing and culture-bound religious texts from other times and other places are not convincing enough to provide paradigms for their personal existence. They require higher

standards of evidence... These children of the Age of Science, myself included, prefer to arrive first-hand, experimentally, at their own conclusions as to the nature and limits of reality. Shamanism provides a way to conduct these personal experiments, for it is a methodology, not a religion.[26]

The immersive methods of phenomenology, co-operative inquiry and ethnography clearly differ in ethos and assumptions from the natural science model of studying human beings. Indeed, in some ways they are opposed. Nevertheless, opposites are often complementary, and in the study of human beings, this seems to be the case. The social sciences are increasingly coming to the conclusion that William James came to a hundred years ago, which is that the hard science quantitative approach and the immersive qualitative approach can be integrated to create a broader understanding of the human condition than either brings alone.[27] This productive mix is illustrated in the study of contemporary spirituality, which has benefited from data gathered through quantitative methods and immersive qualitative methods.

Quantitative surveys with large representative samples of participants have shown that many spiritual beliefs have remained prevalent despite the decline in religious participation, for example, the prevalence of belief in the afterlife has remained constant at around 50% in the UK since the 1930s.[28] From psychometric studies, there is evidence that certain personality characteristics predict being active in spirituality, such as the trait *Openness to Experience*.[29] From longitudinal research, there is evidence that beliefs and practices tend to increase with age.[30] These findings, and many others, have provided a growing picture of how spirituality is distributed and predicted at a broad level.

Thanks to qualitative research, first-person accounts of spiritual experiences are better documented than ever. For example, the Alistair Hardy Religious Experience Research Centre has collected over 6,000 firsthand accounts of spiritual or religious experience.[31] Other qualitative studies of spiritual experiences in various contexts have provided important in-depth insights into what such experiences are *like*, including Peter and Elizabeth Fenwick's work on death-bed visions, and Jenny Wade's research on spiritual experiences during sex.[32,33] Anthropological investigations, such as the ethnographic work of Paul Heelas and Linda Woodhead conducted in the British town of Kendal, have provided important insight into how contemporary non-religious spirituality manifests within Western culture, and how it intersects with the well-being movement.[34]

This brings us finally to the point in the book where it is time to fit the seven paths together within an integrative scheme. Each dialectic has so far been explored on its own terms, but all are expressions of an underlying pattern. In the next chapter, I will show how this pattern can be depicted visually, and how it has parallels with fundamental ideas in mysticism, philosophy and brain science.

Chapter 10

MODI and the Wisdom of the Whole

In the parable of the blind men and the elephant, which originates in India, a group of blind men gather round an elephant and feel it, to learn what it is. Each one touches a different part of the elephant – the ear, side, tusk, trunk, tail, and leg. The one who feels the leg says the elephant is like a pillar, the one who touches the trunk says it is like a branch, the man who holds the tail says it is like a rope, and so on. A sighted man then explains to them that they are disagreeing because they are touching different parts of one animal, so despite their different conclusions, they are all right, but only partially so.

Each of the seven paths that have been presented in this book is akin to one of the blind men touching the proverbial elephant – a partial viewpoint that adds something to the whole. I have discussed them separately to aid their clear presentation, but they are all interrelated parts of a pattern. In this chapter, I endeavor to show them all to be expressions of a fundamental duality, which in turn is also a unified whole. As an aid to presenting this visually, I use the principles of the mandala to depict the *Multiple Overlapping Dialectics* (MODI) model.

A mandala is a geometrical image organized as a series of concentric shapes around a central point. Mandalas have been used across cultures and eras to depict coherence, harmony, wholeness and balance. Examples of mandala images include Tibetan Buddhist healing mandalas, rose windows, Native American medicine wheels, the Celtic cross, Hindu yantras, circular Islam mosaics and the Pagan wheel of the year. Carl Jung studied the use of mandalas in religion, mysticism and dreams, while drawing many himself. He concluded from his research and practice that mandalas are powerful and archetypal depictions of wholeness through the balance of opposites. In his words:

> The mandala is an archetypal image whose occurrence is attested throughout the ages... [It] is probably the simplest model of a concept of wholeness, and one which spontaneously arises in the mind as a representation of the struggle and reconciliation of opposites...[1]

Paralleling the message of the mandala, my objective in this chapter is to depict the wholeness and balance of opposites that emerge from viewing the seven paths together within a singular scheme. I also aim to show that such holism provides for a special kind of wisdom; that which sees all things *sub specie totius* – in light of the whole.[2]

Figure 10.1 presents a visualization of the MODI model of science and spirituality, using the structure of a mandala. With the polarities arranged in this way around a single point, a 14-sided tetradecagon shape emerges. The space within this shape represents the range of human knowing that the open-minded seeker of truth can explore via scientific and spiritual methods.

The left side of the MODI model includes the seven features of science: a focus on outer reality and publicly available facts; an impersonal and detached mode of inquiry; the use of empirical evidence; rational and analytical thinking; mechanistic explanation; knowledge conveyed in numbers and words; and generalized explanation. By keeping its focus on the left-hand side of the diagram as much as possible, science ensures its rigor and replicability. However, science does draw on the features of the right side too, albeit to a lesser degree, for example in the often unspoken feelings and intuitions that motivate scientists (as discussed in Chapter 5).

Spirituality takes the right side of the diagram as its principal abode. It emphasizes the depth of the inner life; personal *I-Thou* experiences with other conscious subjects; the transcendental and mysterious; deep and eudaimonic feeling; forms of nonverbal knowing; ineffable truths that belie language; and the merging of subject and object in contemplation. It draws on much of the left-hand side too to a lesser degree, for example in its verbalized and general theories of spiritual development.

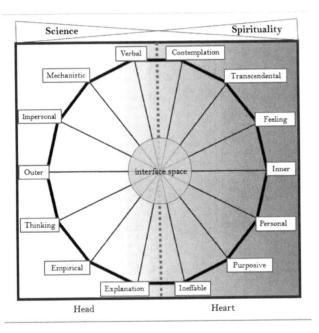

Figure 10.1 The Multiple Overlapping Dialectics (MODI) model of
science and spirituality

The two sides of the figure are separated by a dotted line, which represents the permeable division between head and heart, across which there is constant give-and-take. Around the central point is a circle labeled "interface space." This represents the area where science and spirituality meet and overlap, examples of which have been presented within each of the previous chapters. It is here that we find science-spirituality hybrids such as mindfulness research; consciousness theory; near-death experience studies; scientific theories that speak of purpose and fine-tuning in nature and the universe; sacred geometry; and the study of spiritual experience in the social sciences. This interface space is becoming increasingly populated as science and spirituality seek new ways of cohabiting more closely than in the past.

If an individual explores and cultivates knowledge and experience across both the left and right sides of the MODI model, a more holistic and expansive view of reality is achieved than by pursuing one side alone. The result of this is a special kind of wisdom that comes via the interaction of head and heart.[3]

As a living example of such wisdom, I point the reader to the autobiography of primatologist Jane Goodall, *Reason for Hope: A Spiritual Journey*.[4] Her work exemplifies balancing of the left side and right side of the MODI model. In her scientific study of primates, she is rigorous, analytical and as objective as she can be, while in her mystical and spiritual life she feels a deep, numinous connection with the animals that she studies and the natural world as a whole. She describes her life as being a balance of these two sides. She is currently in her 80s and still touring the world doing talks on the ethical imperative of integrating scientific and spiritual modes of being, and what we must do to avoid an impending ecological collapse.

As another contemporary example of head-heart wisdom, the current Dalai Lama effectively balances these through cultivating spiritual and scientific awareness. This is exemplified in his work in co-founding the *Mind and Life Institute*, which has forged important links and dialogues between science and Buddhist spirituality.

The idea that wisdom is the product of dialectical balance between head and heart has a heritage in the East within yin-yang philosophy and the West in the tradition of alchemy. In yin-yang philosophy, the balance of yin and yang in character and lifestyle is said to facilitate the development of wisdom. Yang relates very clearly to the left of the MODI model – it is knowledge based on rationality, assertion and dominance, explicit propositions, visible things, solid objects and external reality. It is represented by the metaphors of light and sun, and by masculinity. In contrast, yin equates to the right side of the MODI model – it relates to knowledge based on feelings, intuition, ineffable gnosis, the unconscious, and being receptive to transcendental or hidden influence. It is represented by images of shade, night and the moon, and by femininity.[5]

An enduring visualization of yin and yang is the taijitu (see Figure 10.2). Yin is the black side; yang is the white side. The dots represent how each side contains its opposite. The shape can be viewed as one circle or two teardrop shapes, depending on how you look at it. It is thus *both* one *and* two, and there is no contradiction in it being both. The singular wholeness of the image represents the primal oneness of reality, variously called *taiji* or *wuji*.

Figure 10.2 The Taijitu: A visualization of yin-yang as both two and one

The wise person, manifesting the higher oneness of yin-yang, avoids all extremes and seeks virtue in countering too much of one thing with its opposite. They become manifestations of the taijitu – a whole composed of the balance of twoness, which in turn shows as a combination of masculine and feminine properties.

The duality of the MODI model also has parallels with the distinction in alchemy between two archetypal principles called *Sol* and *Luna*. Alchemy is a Western wisdom tradition that focuses on the processes of *transformation*. It was a major influence on the work of Carl Jung, and via him on contemporary spirituality.[6] In its study of metals, alchemy explores how to transform base metals into gold. In its spiritual aspect, it considers how the human personality can transform into a higher unity. To pursue the alchemical path, the practitioner has to integrate opposites, the most important and all-encompassing of which are Sol and Luna. Sol refers to rational, thinking knowledge, external focus and methodical planning, and reflects the left side of the MODI model. It is represented by the images of the sun and a King who sits over a cave containing a dragon. Luna refers to mystical knowing, feelings and unconscious influence, and so links to the right side of the MODI model. She is represented by the Queen, the moon and the sea.

Luna is, according to alchemy, our original and unconscious nature, and Sol is our learned and conscious nature. Sol seeks to subdue Luna in the early stages of development, but higher conscious integration entails a working harmony or *sacred marriage* between these two archetypes. As the alchemist goes through a transformation from fragmented personality to higher spiritual being, he or she must learn how to balance these two forces of existence and their manifestations as thought and feeling. For a full description of the symbolism of Sol and Luna within the context of an alchemist's mandala, the reader is directed to the essay by Dennis William Hauck in the following endnote.[7]

Both yin-yang philosophy and alchemy concur that the aim of development is not merely to conjoin opposites but to transcend them and move *beyond* opposites. Jung, a lifelong student of alchemical symbols and ideas, used the term *complexio oppositorum* (the complex of opposites) to describe how the divine transcends yet includes opposites. He refers to the primal reality that lies beyond dualities as the *unus mundus*. He suggested that words such as God, Brahman or Nirvana are all terms that point towards the ultimate unity that somehow contains the duality and multiplicity of life in a mysterious *both-and* embrace.

This experience of transcending opposites is frequently referred to as *nonduality*. According to those who experience it, the nondual state of consciousness brings a profound sense of deep truth, love and insight, while paradoxically transcending all these concepts that imply an opposite. In the nondual state of consciousness, all divisions and oppositions are shown to be nothing more than the shadow play of the thinking mind. All the dialectics in the MODI model are thus paths that lead beyond themselves to the same unified ground. This experience of being one with the ultimate whole that contains all of reality's many parts, is one that the analytical Western mentality struggles with. That may in part be explicable by an imbalanced use of our brains.

MODI and the brain

Most vertebrate species including mammals and birds have a brain that is structured as two lateral hemispheres split down the middle by a fissure. The two hemispheres are connected by a single bundle of fibers called the corpus callosum. They look similar to the untrained eye, but in fact, they differ in weight, shape, cell structure, grey-to-white matter ratio, neuroendocrine responsiveness and neurotransmitter profile. For a century or more, researchers have explored the difference between the two hemispheres in terms of how they relate to attention, memory, thought, perception and emotions. Iain McGilchrist has recently integrated and synthesized this research in his work *The Master and his Emissary*.[8]

McGilchrist outlines key functional differences between the two hemispheres. In terms of attention, the left hemisphere focuses on parts and separate objects, while the right hemisphere has a global form of attention that attends to wholes, patterns and connections. In terms of perception, the left hemisphere sees things through labels, concepts and categories, while the right hemisphere sees connected systems and wholes composed of interconnected elements. The left hemisphere attends to whatever is inanimate, mechanical, impersonal or machine-like, while the right hemisphere is personal, flexible, organic and empathic. In relation to music, the left hemisphere is sensitive to rhythm, while the right comprehends melody. In terms of emotion, anger is served by the left hemisphere, while the right deals with recognition and production of most emotions and the experience of compassion. There is give-and-take between the functionally different hemispheres but there is substantial independence too. Indeed a person can function with just one, as was shown in the case of a healthy 10-year-old girl who was born with only one hemisphere.[9]

Of all the differences between the hemispheres, McGilchrist considers the difference in attentional focus to be a key to understanding the rest. The left hemisphere attends to parts, the right hemisphere to wholes. An example research study helps to highlight this difference. David Navon at the University of Haifa asked brain-damaged patients to copy a picture that showed 20 small copies of the letter A arranged into the shape of a large H. Patients with damage to the left hemisphere, who were reliant on their right hemisphere, generally could see the overall pattern but could not see the parts, so drew the large letter H with no small A letters included. In contrast, patients with damage to the right hemisphere could only see the parts but not the whole pattern, hence drew small A letters all over the page with no H pattern.[10]

There are clear parallels between the functional roles of the two hemispheres and the two sides of the MODI model. Both the left side of the brain and the science-based left side of the MODI model are focused on impersonal, abstract, uninvolved, unemotional, factual information, and tend to be reductionist and mechanistic. In contrast, both the right hemisphere of the brain and the spirituality-oriented right side of the MODI model are more personal, particular-focused, holistic, emotional and aesthetic. Fitting with this, McGilchrist has written that he believes the right hemisphere is more attuned to spirituality:

> The right hemisphere is tolerant of ambiguity and paradox, and understands the coming together of opposites… For these reasons, it is perhaps more open to the spiritual than the left hemisphere, which is always trying to reduce experience to something else it already 'knows'.[11]

McGilchrist argues that the evolution of the brain's hemispheric split must have a clear functional purpose, for it is a physiologically more complex solution than a non-lateralized brain. He theorizes that it is a physiological representation of two fundamental ways of experiencing the world, and of keeping these relatively separate. Facilitated by the right hemisphere and exemplified in humans by spirituality, the *holistic* way of knowing experiences the world as a seamless whole in flux, without the discrete labels of language and rationality. In contrast, the *atomistic* way of knowing and paying attention divides the world up into parts, conceptualizes it into abstractions, and focuses narrowly on particular parts. The holistic and atomistic ways of knowing are so fundamental to life that the brain provides a functional basis for their expression in all vertebrates. This allows the brain to have a specific beam of attention on a particular object or issue, while also having a diffuse global awareness of the environment as a whole, in case another threat or opportunity arises.

The atomistic thinking that the left hemisphere specializes in can easily become ignorant of all the holistic and more ineffable work done by the right. McGilchrist suggests that we in the West have forgotten the give-and-take of whole and part, and that the left hemisphere has run amok with its dislocated rationality. Western culture, he says, overly values the rational, machine-like and mechanical, and tends to see narrow issues instead of the big picture. Philosophy and science that is excessively dominated by the left hemisphere, therefore, tends to assume that *everything* can be known as measurable objects through the filter of reason, and that any mysterious reality beyond the known is impossible. This kind of left-hemisphere dominated viewpoint is represented within philosophy as the doctrine of physicalism.

MODI and philosophy: Beyond physicalism

The doctrine of physicalism asserts that reality is comprised solely of matter, energy and the fundamental forces of physics. It further posits that mind and consciousness exist only in the brains of certain complex animals, as 'epiphenomenal' products of brain activity that do not actually cause or affect anything. Our sense that consciousness does make a difference, for example that a consciously-made decision leads to personal responsibility for outcomes, is an illusion. Really all that is going on is brain activity leading to body movements. A metaphor that is sometimes used for this is that the conscious mind is like steam coming off the engine of the brain. Given that the experiences of having purposes and making free decisions are illusory in physicalism, the doctrine entails that all notions of human agency and free will must be jettisoned. We are determined not by our choices, but by our atoms and the laws that govern them.

In relation to the MODI model, adhering to the philosophy of physicalism means interpreting it in such a way that the left side – being outwardly focused on the material substance and mechanisms of things – gathers information that is more accurate and real than the right side, with its inner focus on subjectivity and intent.

Many maintain a place for spirituality in life while professing physicalism, for example the scientist Carl Sagan, whose ideas about spirituality were discussed at the end of Chapter 5. Nevertheless, while spirituality can be squeezed into a physicalist worldview as a kind of enjoyable illusion, there are problems with physicalism as a metaphysic that have been mentioned in previous chapters and are worth reiterating in summary here. Firstly, physicalism has no explanation as yet for how the subjective richness of consciousness could actually emerge from the matter of the brain. As science currently stands, the process of conjuring an interior, conscious point of view from the inert matter of the brain is a miracle that arguably surpasses turning water into wine. It's not that non-physicalist theories of consciousness (such as Integrated Information Theory, discussed in Chapter 4) provide a perfect explanation as yet, but until there is a conclusive answer either way, all options should be pursued, including first-person approaches that are based on meditation and altered states of consciousness.

Another quandary is that physicalism has no agreed answer as to why the structure of the universe accords to the rules of reason and the highly abstract language of equations. The elegant logico-mathematical architecture that scientists find underlying reality may be a fluke or a projection of our own minds, but it may equally be the mark of cosmic intelligence, or even a sign that the universe is a simulation that has been designed by hyper-intelligent beings (as discussed in Chapter 7). One can of course reasonably doubt the presence of higher intelligence or purpose, on the basis of the small portion of the universe that can be seen and measured, but this is not a logical necessity.

Philosophies that assume mind and consciousness to be

inherent to the cosmos are not just peripheral or fringe ideas in science, but have been adopted by many influential theorists. For example, of the pioneering quantum physicists Erwin Schrodinger, Wolfgang Pauli, Arthur Eddington and Max Planck, all of them adopted non-physicalist metaphysics. Schrodinger based his work on the mystical philosophy of Vedanta, which sees individual consciousness as an expression of a unitary cosmic consciousness. Pauli developed a form of dual-aspect monism in partnership with Carl Jung that sees mind as present at all levels of reality.[12] Eddington was a Quaker and adopted the philosophy of idealism in which mind is primary, and matter secondary. Max Planck also openly stated his view that consciousness is fundamental to reality.[13] The Copenhagen Interpretation of quantum theory drew on these philosophies, by suggesting that reality as we know it – the world of objects in space and time – comes into being within the consciousness of an observer.[14]

Over the course of this book, I have presented a number of philosophies that describe a reality in which mind and/ or consciousness are inherent to the cosmos as a whole. These include the animism of Thomas Berry and Emma Restall Orr; Ken Wilber's four-quadrant integral theory; the evolutionary pantheism of Schelling and Hegel; the panpsychism of Christof Koch; the dual-aspect monism of Jung and Schopenhauer; the transcendentalism of Ralph Waldo Emerson and Margaret Fuller; the archetypal cosmology of Richard Tarnas; the biocentrism of Robert Lanza; the organicism of Rupert Sheldrake; the Neoplatonic philosophy of Huston Smith; and the transpersonal theories of Stanislav Grof and Charles Tart. Viewed from any of these non-physicalist perspectives, spirituality is a mode of inquiry that captures those mental, intentional and interior aspects of primary reality that science does not. This gives spirituality an equal partnership with science, and this in turn provides support for the dialectical 'both-and' ethos of the

MODI model.

As a way of staying open to the mystery that lies beyond all these philosophies, it is helpful to reflect that science and spirituality are recent entrants on the human scene, relative to the 200,000 years that homo sapiens has been around. As recognized domains of inquiry away from philosophy and religion, they have had about 400 years of existence. If we were to condense human history into a year, this period would equate to about fifteen hours. As for the future, what we will know two or three millennia down the line will be as improbable and outlandish to our contemporary minds as the Internet and quantum entanglement would be to our hunter-gatherer ancestors. So let's hold our cherished ideas lightly, as they will all be surpassed in time. New wonders lie in store.

Epilogue

The Interconnected Age

We are living through a revolution in connectedness, which is often as exhausting as it is exhilarating. With every passing year, information, knowledge, goods, and people move around the world in ever faster, more voluminous and more complex networks. It is hard to keep up. From 1980 to the present day, international trade has increased by over 500 percent, aviation by 400 percent, and the number of computers on the Internet has increased from zero to three billion.[1]

We are learning how to exist as nodes within this global informational nexus; as beings without clear boundaries who can have global effects at the click of a button or the touch of a screen. This is a psychological challenge without precedent in human history, and many people are struggling to keep up. The burnout from information overload occasioned by e-mail, social media and 24-hour news is shown in epidemic levels of mental health problems across the Western world, particularly in teenagers and young adults.[2,3] This rise in mental illness has been linked to the stressors of social media and the Internet.[4]

Interconnectedness brings other anxieties too, for example, it exposes us to the nefarious intents of criminals from all across the world. A ransomware virus recently crippled computers across 120 countries in a matter of hours, asking for money to unlock files. Such hazards are inherent to the interconnected world and the technologies that underpin it.

One response to these new dangers is to step back from interconnectedness and retrench into strong nationalistic or religious identities. This reaction can be seen in the current rise of extreme right-wing politics and fundamentalist religion, both of which are premised on the importance of barriers, fear of other cultures, and reducing the open flow of people and information across the world. These regressive reactions to the change are understandable as many people long for the less complicated locality of the past. Yet we must stay the course. Interconnection and interdependence bring the genuine possibility of being more united as a species and more able to enact initiatives that serve the needs of the planet as a whole, such as responses to climate change, species extinction, and rainforest destruction.

Science and spirituality provide crucial resources for the head and the heart to be willing parts of the interconnected nexus. Science provides a strong intellectual basis for this. The crux of the scientific worldview, ever since Newton, is the *total interconnection of all phenomena*. Newton found that the tides, the motions of the planets and the fall of an apple were connected by the operation of just one force – gravity. Gravity in Newtonian physics extends out infinitely, so *everything* is subtly pulled by everything else within an infinitely complex tapestry of influence.

Atomic theory also adds to this picture of interconnectedness. Imagine a beach – thanks to atomic theory we know that the sand, cliffs, air and water are all made of a mix of electrons, neutrons and protons, and that particles are exchanged between them constantly. We know that through their mutual interaction, they *have formed each other* over millions of years: The waves turn the rock of the cliffs into sand, the angle of the sand and the wind create the waves, rock is slowly formed on the sea bed, and so on, in webs of complex interconnection.

Quantum physics has now painted an even more connected picture, whereby all things at a subatomic level are composed of superimposed potentialities that are not clearly located in space and time, called wave functions, the 'collapse' of which into actual particles and material stuff depends on the observer.

So subject and object are connected too. Reality is not a separable thing 'out there', but co-created by us as observers.

A strong knowledge of science is increasingly important given that as individuals we are far more dependent on it than we ever have been. A century ago, a person in an industrialized country could pass their day barely in contact with the products of science. With the advent of the smartphone and wearable technologies, we are now barely ever away from them. This ubiquity of technology can cocoon us from nature and ourselves, and we must make efforts to retain a connection with the world around us and our own bodies. This means educating ourselves in the theories and facts of science where possible, in order to understand the solar system, the natural world, the seasons, our bodies, our food and more. It means also making efforts to understand how the technologies that shape our lives work, such that we feel more like an active participant, rather than a passive pawn, in the high-tech matrix that is contemporary life.

Another quality of the scientific attitude that is crucial for interconnected living is critical thinking. In our world of ubiquitous online information, the validity and accuracy of the facts that reach us is constantly questionable, and so more than ever it is *you* who must filter the truthful from the fallacious, and you need your critical filter working well for the task.

To become acquainted with science is to *know* interconnectedness, while to practice spirituality is to *feel* connectedness, and to live it as an ethos and ethic. All practices that lead to the dissolution of the grasping ego, whether quiet meditation, ecstatic dance or simple acts of kindness, bring about a sense of being deeply enmeshed within a greater whole. Compassion and love emerge spontaneously from this sense of oneness.

Spirituality as the open pursuit of such experiences is well adapted to our world of permeable boundaries. It is a place of expansive possibility and critical inquiry that everyone can explore and journey through. It provides a common ground for individuals of different cultures, nationalities, religions or worldviews to come together in a spirit of openness and quest. Religion, with its messages of continuity and social solidarity, has its place too in the interconnected age, as long as it accepts acting as a center of gravity in a person's life, from which they can explore and return, rather than a closed system of belief and ethics that claims unique deliverance.

With the constant risk of cognitive and affective overload in the connected world, meditative and contemplative practices that bring the mind back to its still center have never been more important for health and sanity. Furthermore, spiritual practices that focus on the body, such as yoga and tai chi, provide a counterbalance to the disembodiment of screen-mediated living and online relationships. The concept of retreat has also never been more important as a way of giving one's mind and body time out from the stresses of everyday life. We all need to spend regular time off the grid.

Our current time will, I believe, be reflected on by historians as the transition out of the modern era.[5] The modern era was all about individual empowerment, and overcoming the herd mentality of traditionalist culture. This trans-modern shift has already moved through the phase of postmodernism, which as a critique of modernism was meant to deconstruct some of its more problematic manifestations, but not really build anything new. What comes next will be about finding a form of integration and wholeness that maintains the hard-won individual freedoms and human rights of the modern era but allows for a greater sense of oneness too.

This new level of cultural and personal wholeness can only be achieved if the split between head and heart is overcome, and a dialectical give-and-take between science and spirituality becomes the norm. Given the opportunities and dangers of the interconnected world, this integration of intellect and love is a developmental task that has never been more pressing. It is an essential and urgent challenge for our time.

Endnotes

Chapter 1: Setting the Scene

1. Einstein, A. (1936) "Physics and Reality." Translation by Jean Piccard. *Journal of the Franklin Institute, 221,* 349–382.
2. Armstrong, K. (2007) *The Great Transformation: The World in the Time of Buddha, Socrates, Confucius and Jeremiah.* Atlantic Books.
3. A notable example of the difficulties in arguing whether religion has caused more positives than negatives, or vice versa, was illustrated by the debate on the topic between Christopher Hitchens and Tony Blair. Both sides offered different kinds of evidence and could not agree on their relative merits. See: https://www.youtube.com/watch?v=xFnSjmQCGDM
4. Foster Jones, R. (1961) *Ancient and Moderns: A Study of the Rise of the Scientific Movement in Seventeenth-Century England.* Dover Publications.
5. Wulff, D. (1996) *Psychology of Religion: Classic and Contemporary Views.* New York: John Wiley and Sons.
6. Rousseau, D. (2014) "A Systems Model of Spirituality." *Zygon, 49,* 476–507.
7. Roof, WC (1999) *Spiritual Marketplace: Baby Boomers and the Remaking of American Religion.* Princeton University Press.
8. Fuller, RC (2001) *Spiritual but not Religious: Understanding Unchurched America.* Oxford University Press.
9. Ferrer, J. (2011) "Participatory Spirituality and Transpersonal Theory: A Ten-Year Retrospective." *The Journal of Transpersonal Psychology, 2011, 43,* 1–34.
10. Ammerman, NT (2013) "Spiritual But Not Religious? Beyond Binary Choices in the Study of Religion." *Journal for the Scientific Study of Religion, 52,* 258–278.
11. Ravindra, R. (2001) *Science and the Sacred.* Theosophical

Publishing House.
12. Rowson, J. (2014) *Spiritualise: Revitalising spirituality to address 21ˢᵗ century challenges*. RSA Publications.
13. Bohannon, J. (2013) "Who's Afraid of Peer Review?" *Science*, 342, 60–65.
14. Gottlieb, RS (2012) *Spirituality: What It Is and Why It Matters*. Oxford University Press.
15. Damasio, A. (2006) *Descartes' Error: Emotion, Reason and the Human Brain*. Vintage Books.
16. The poll was of 44,000 people from 30 countries. See: http://www.theguardian.com/science/alexs-adventures-in-numberland/2014/apr/08/seven-worlds-favourite-number-online-survey
17. Gould, SJ (1999) *Rocks of Ages: Science and Religion in the Fullness of Life*. London: Jonathan Cape.
18. Xinyan, J. (2013) "Chinese Dialectical Thinking – The Yin Yang Model." *Philosophy Compass*, 8/5, 438–446.
19. *ibid*.
20. Capra, F. (1991) *The Tao of Physics: An Exploration of the Parallels between Modern Physics and Eastern Mysticism*. Flamingo.
21. Gier, N. (1983) "Dialectic: East and West." *Indian Philosophical Quarterly* 10 (January, 1983), pp. 207–218.
22. Robinson, OC (2015) "Emerging adulthood, early adulthood, and quarter-life crisis: Updating Erikson for the twenty-first century." In R. Žukauskiene (ed.), *Emerging Adulthood in a European Context* (pp. 17–30). New York: Routledge.
23. Basseches, M. (1984) *Dialectical Thinking and Adult Development*. Praeger.
24. Basseches, M. (2005) "The Development of Dialectical Thinking As An Approach to Integration." *Integral Review*, 1, 47–63.
25. Kegan, R., Noam, GG & Rogers, L. (1982) "The psychologic of emotion: A neo-Piagetian view." In D. Cicchetti & P. Hesse

(eds.), *New Directions for Child and Adolescent Development, no.16.* San Francisco: Jossey Bass.

26. Mansfield, V. (2002) *Head and Heart: A Personal Exploration of Science and the Sacred.* Quest Books.

Chapter 2: Entangled Histories

1. Johansson, SR. "Medics, Monarchs And Mortality, 1600–1800: Origins Of The Knowledge-Driven Health Transition In Europe." *University of Oxford Discussion Papers in Economic and Social History,* Number 85, October 2010. http://www.nuff.ox.ac.uk/economics/history/Paper85/johansson85.pdf

2. Foster Jones, R. (1961) *Ancient and Moderns: A Study of the Rise of the Scientific Movement in Seventeenth-Century England.* Dover Publications.

3. *ibid.*

4. *ibid*, p. 10 and p. 4.

5. Blanning, T. (2011) *The Romantic Revolution.* London: Phoenix.

6. Tarnas, R. (1996) *The Passion of the Western Mind: Understanding the Ideas That Have Shaped Our Worldview,* p. 368. Pimlico.

7. Mill, JS (2006) *On Liberty and the Subjection of Women,* p. 132. Penguin Classics.

8. Blanning, T. (2009) *The Triumph of Music: Composers, Musicians and Their Audiences, 1700 to the Present.* London: Penguin.

9. Blanning, T., *ibid.*

10. Naumann, W. (1952) "Goethe's Religion." *Journal of the History of Ideas, 13,* No. 2, pp. 188–199.

11. Quoted in Ferber, M. (2010) *Romanticism: A Very Short Introduction,* p. 54. Oxford University Press.

12. Holmes, R. (2009) *The Age of Wonder: How the Romantic Generation Discovered the Beauty and Terror of Science.* London: HarperPress.

13. Razavy, M. (2002) *Quantum Theory of Tunneling.* World

Scientific Publishing Co.

14. *The Sacred Books of the East,* a 50-volume set of English translations of Asian religious writings, edited by Max Müller and published by Oxford University Press, which were published between 1879 and 1910, were key to this.

15. Schmidt, LE (2012) *Restless Souls: The Making of American Spirituality.* University of California Press.

16. Auricchio, L. (2004) "The Nabis and Decorative Painting." In *Heilbrunn Timeline of Art History.* New York: The Metropolitan Museum of Art. Retrieved from: http://www.metmuseum.org/toah/hd/dcpt/hd_dcpt.htm

17. Eller, Cynthia (1993) *Living in the Lap of the Goddess: The Feminist Spirituality Movement in America.* New York: Crossroad.

18. Adler, F. (1905) *The Essentials of Spirituality.* New York: James Pott & Co.

19. Vaillant, G. (2008) *Spiritual Evolution: A Scientific Defense of Faith.* New York: Broadway Books.

20. Einasto, J. (2010) "Dark matter." In *Astronomy and Astrophysics.* In Oddbjorn Engvold, Rolf Stabell, Bozena Czerny, John Lattanzio (eds.), *Encyclopedia of Life Support Systems (EOLSS),* p. 152.

21. Moskowitz, C. (2011) *What's 96 Percent of the Universe Made of? Astronomers Don't Know.* Space.com, http://www.space.com/11642-dark-matter-dark-energy-4-percent-universe-panek.html

22. Tacey, D. (2004) *The Spirituality Revolution: The Emergence of Contemporary Spirituality.* Routledge.

23. Heelas, P. & Woodhead, L. (2005) *The Spiritual Revolution: Why Religion is Giving Way to Spirituality.* Blackwell Publishing.

24. Michaud, D. (2012) "Religious and mystical experiences common among Americans." http://www.patheos.com/blogs/scienceonreligion/2012/02/religious-and-mystical-

experiences-common-among-americans/
25. Fuller, RC (2001) *Spiritual, but not Religious: Understanding Unchurched America.* Oxford: Oxford University Press.
26. King, M., Marston, L., McManus, S., Brugha, T., Meltzer, H., Bebbington, P. (2013) "Religion, spirituality and mental health: results from a national study of English households." *British Journal of Psychiatry, 202,* 68–73.
27. Saucier, G. and Skrzypinska, K. (2006) "Spiritual But Not Religious? Evidence for Two Independent Dispositions." *Journal of Personality, 74,* 1257–92.
28. Tarnas, R. (2013) "The role of 'heroic' communities in the postmodern era." *Digital Commons.* http://digitalcommons.ciis.edu/founderssymposium/30/

Chapter 3: Outer – Inner

1. Jung, CG (1962/1995) *Memories, Dreams, Reflections.* Flamingo.
2. Galilei, Galileo (1623) *The Assayer* (Italian: *Il Saggiatore*).
3. Lanza, R. (2010) *Biocentrism: How Life and Consciousness are the Keys to Understanding the True Nature of the Universe.* BenBella Publishers.
4. Darwin, C. (1839/1905) *The Voyage of the Beagle.* Freeman. Wallace, AR (1869) *The Malay Archipelago.* Macmillan and Co.
5. NOAA (2014) "How much of the ocean have we explored?" http://oceanservice.noaa.gov/facts/exploration.html
6. Colitt, R. (2007) "Brazil sees traces of more isolated tribes." http://uk.reuters.com/article/uk-brazil-amazon-indians-idUKN1728525620070117
7. CBC News: Technology and Science (2014) "Uncontacted Amazon tribe meets modern world in Brazil." http://www.cbc.ca/news/technology/uncontacted-amazon-tribe-meets-modern-world-in-brazil-1.2704427
8. Sagan, C. (1997) *Pale Blue Dot: A Vision of the Human Future in*

Space, 1997 reprint, pp. xv–xvi.

9. Swimme, B. (1999) *The Hidden Heart of the Cosmos: Humanity and the New Story*. Orbis Books, p. 52.

10. Chivian, E. and A. Bernstein (eds.) (2008) *Sustaining Life: How Human Health Depends on Biodiversity*. Center for Health and the Global Environment. New York: Oxford University Press.

11. Thomas, CD et al. (2004) "Extinction risk from climate change." *Nature* 427: 145–148.

12. Watson, JB (1913) "Psychology as the Behaviorist Views it." *Psychological Review, 20*, 158–177.

13. Bretherton, I. (1992) "The origins of attachment theory: John Bowlby and Mary Ainsworth." *Developmental Psychology, 28*, 759–775.

14. Open Science Collaboration (2015) "Estimating the reproducibility of psychological science." *Science, 349, Issue 6251*. DOI: 10.1126/science.aac4716.

15. Locke, EA & Latham, GP (1990) *A Theory of Goal Setting and Task Performance*. Prentice-Hall.

16. Watson, R. (1856) *Watson's Biblical and Theological Dictionary*. Carlton and Porter.

17. Milton, J. (1667) *Paradise Lost, Book I*. London: Samuel Simmons.

18. Schmidt, LE (2005) *Restless Souls: The Making of American Spirituality*, pp. 110–111. HarperSanFrancisco.

19. Clarke, TC et al. (2015) "Trends in the use of complementary health approaches among adults: United States, 2002–2012." *National health statistics reports*; no 79. Hyattsville, MD: National Center for Health Statistics. https://nccih.nih.gov/research/statistics/NHIS/2012/mind-body/meditation

20. McDonald, K. (2005) *How to Meditate: A Practical Guide*. Wisdom Publications.

21. *Surangama Sutra*.

22. Kornfield, J. (1989) "Obstacles and Vicissitudes in

Spiritual Practice." In *Spiritual Emergency: When Personal Transformation Becomes a Crisis* (eds. Grof, S. and Grof, C.), Tarcher.

23. Kavanagh, J. (2007) *The World is Our Cloister: A guide to the modern religious life*. O-Books.

24. McDonald, K., *ibid*.

25. Huxley, A. (2009) *The Perennial Philosophy*, p. 89. HarperPerennial.

26. Russell, Peter (2002) *From Science to God*.

27. Smith, H. (1976) *Forgotten Truth: The Common Vision of the World's Religions*, p. 75. HarperOne.

28. Brunton, P. (2016) *The Inner Reality*, p. 26. Eighth edition. London: Rider and Company.

29. Assagioli, R. (1980) "Self-realization and Psychological Disturbances." In *Spiritual Emergency: When Personal Transformation Becomes a Crisis* (eds. Grof, S. and Grof, C.), Tarcher.

30. Tolle, E. (2001) *The Power of Now: A Guide to Spiritual Enlightenment*, p. 109. Yellow Kite.

31. Harris, S. (2015) *Waking Up: Searching for Spirituality without Religion*, p. 140. Black Swan. (Italics added.)

32. Hutcherson, CA, Seppala, EM, Gross, JJ (2008) "Loving-Kindness Meditation Increases Social Connectedness." *Emotion, 8*, 720–724.

33. Tacey, D. (2004) *The Spirituality Revolution: The Emergence of Contemporary Spirituality*, p. 41. Routledge.

34. Kabat-Zinn, J. (1982) "An out-patient program in behavioral medicine for chronic pain patients based on the practice of mindfulness meditation: theoretical considerations and preliminary results." *General Hospital Psychiatry, 4*, 33–47.

35. Nichols, AL & Maner, JK (2008) "The Good Subject Effect: Investigating Participant Demand Characteristics." *Journal of General Psychology, 135*, 151–165.

36. Fox, K. et al. (2014) "Is meditation associated with altered

brain structure? A systematic review and meta-analysis of morphometric neuroimaging in meditation practitioners." *Neuroscience & Biobehavioral Reviews, 43,* 48–73.

37. Dobkin, PL, Irving, JA, Amar, S. (2012) "For Whom May Participation in a Mindfulness-Based Stress Reduction Program be Contraindicated?" *Mindfulness 3,* 44–50.

38. Purser, R. & Loy, D. (2013) "Beyond McMindfulness." *Huffington Post.*

39. McDonald, K. (2005), *ibid.*

40. Grof, S. and Grof, C. (1989) "Spiritual Emergency: Understanding Evolutionary Crisis." In Grof, S. and Grof, C. (eds.) *Spiritual Emergency: When Personal Transformation Becomes a Crisis.* New York: Tarcher.

41. Laing, RD (1989) "Transcendental Experience in Relation to Religion and Psychosis." In Grof, S. and Grof, C. (eds.) *Spiritual Emergency: When Personal Transformation Becomes a Crisis.* New York: Tarcher.

42. Jung, CG (1978) *Man and His Symbols.* Picador.

43. Dix, M. (2014) "What is the dark side of meditation?" Retrieved from: http://aboutmeditation.com/the-dark-side-of-meditation

44. Bly, R. (1988) *A Little Book on the Human Shadow.* HarperCollins.

45. Epstein, M. (1998) "Therapy and meditation." *Psychology Today, 31,* 46–53. Retrieved from: http://enlight.lib.ntu.edu.tw/FULLTEXT/JR-ADM/mark.htm

46. Vaughan, F. (2000) *The Inward Arc: Healing in Psychotherapy and Spirituality.* Lincoln, NE: Backinprint.com.

47. Johnson, RA (1986) *Inner Work: Using Dreams & Active Imagination for Personal Growth.* New York: HarperOne.

48. Johnson, RA, *ibid.*

49. Jung, CG & Pauli, W. (1952) *The Interpretation of Nature and the Psyche.* Ishi Press.

50. Schopenhauer, A. (1819/1995) *The World as Will and Idea*, p. 5.

Phoenix.

51. Jung, CG (1963) *Mysterium Coniunctionis: An Inquiry into the Separation and Synthesis of Psychic Opposites in Alchemy.* Routledge.

52. Wilber, K. (2000) *The Marriage of Sense and Soul.* Boston: Shambhala Publications.

Chapter 4: Impersonal – Personal

1. Berry, T. (1988) *The Dream of the Earth,* pp. 14–15. San Francisco: Sierra Club Books.

2. Buber, M. (2000) *I and Thou.* Scribner Classics.

3. Solomon, RC (2006) *Spirituality for the Skeptic: The Thoughtful Love of Life,* p. 68. Oxford University Press.

4. Nagel, T. (1989) *The View from Nowhere.* Oxford University Press.

5. Taylor, SE (1991) *Positive Illusions: Creative Self-deception and the Healthy Mind.* Basic Books.

6. Rammstedt, B. & John, OP (2007) "Measuring personality in one minute or less: A 10-item short version of the Big Five Inventory in English and German." *Journal of Research in Personality, 41,* 203–212.

7. Goethe, JWV (1792) *The Experiment as Mediator between Subject and Object.* Retrieved from: http://v1.elfieraymond.com/bestfoot.html

8. Pasteur, L. (1870) *Etudes sur la maladie des vers á soie,* p. 39.

9. Biber, D. & Gray, B. (2013) "Nominalizing the verb phrase in academic science writing." In (eds. B. Aarts et al.) *The Verb Phrase in English: Investigating Recent Language Change with Corpora* (pp. 99–132). Cambridge: Cambridge University Press.

10. Kline, M. (2003) *Mathematics and the Physical World.* Dover Publications.

11. Tutu, D. (1999) *No Future Without Forgiveness.* Rider.

12. Rumi, J. & Cowan, J. (1997) "Kulliyat-e Shams, 2114." In

Rumi's Divan of Shems of Tabriz: A New Interpretation by James Cowan. Element Books.

13. Nagel, Thomas (1974) "What is it like to be a bat?" *The Philosophical Review*, 83, 435–450.

14. Underhill, E. (1911/2016) *Mysticism: A Study in the Nature and Development of Spiritual Consciousness*. Ancient Wisdom Publications.

15. de Chardin, T. (1955) *The Phenomenon of Man*, p. 267. Collins.

16. Frankl, VE (2011) *Man's Search for Ultimate Meaning*. London: Rider.

17. Fyfe, A. (1951) *Dances of Germany*. London: Max Parrish.

18. Mercadante, LA (2014) *Belief without Borders: Inside the Minds of the Spiritual but not Religious*. Oxford University Press.

19. Abram, D. (1997) *The Spell of the Sensuous: Perception and Language in a More-Than-Human World*. Vintage Books.

20. Neumann, Erich (1991) *The Great Mother: An Analysis of the Archetype*. Princeton, NJ: Princeton University Press.

21. Abram, D., *ibid*.

22. Harvey, G. (2005) *Animism: Respecting the Living World*. C. Hurst & Co Publishers Ltd.

23. Emerson, RW (1982) *Nature and Other Essays*, pp. 38–39. Penguin.

24. Starhawk (1999) *The Spiral Dance: A Rebirth of the Ancient Religion of the Great Goddess*, p. 12. HarperSanFrancisco.

25. Koch, C. (2012) *Consciousness: Confessions of a Romantic Reductionist*, p. 132. MIT Press.

26. Koch, C. (2014) "Is Consciousness Universal? Panpsychism, the ancient doctrine that consciousness is universal, offers some lessons in how to think about subjective experience today." *Scientific American*. https://www.scientificamerican.com/article/is-consciousness-universal/

27. Koch, C., *ibid*.

28. Orr, ER (2012) *The Wakeful World: Animism, Mind and the Self in Nature*. Moon Books.

29. Harris, S. (2015) *Waking Up: Searching for Spirituality without Religion*. Black Swan.
30. Starhawk, *ibid*, p. 220.
31. Berry, *ibid*, p. 14.
32. Sri, PS (n.d.) *Shaw (1856–1950), Gandhi (1869–1948) and Vegetarianism*. http://www.shawsociety.org/Sri.htm
33. Goodall, J. (2004) *Reason for Hope: A Spiritual Journey*. Grand Central Publishing.
34. Home Office (2004) "Statistics of Scientific Procedures on Living Animals, Great Britain 2004." Retrieved from https://www.gov.uk/government/uploads/system/uploads/attachment_data/file/272232/6713.pdf
35. Singer, P. (1995) *Animal Liberation*. Pimlico.
36. Singer, P., *ibid*, p. 38.
37. Seligman, MEP & Maier, SF (1968) "Alleviation of learned helplessness in the dog." *Journal of Abnormal Psychology*, 73, 256–262.
38. Singer, P., *ibid*.
39. *New York Times* (1982) "Value of 2 dolphins set free in '77 at issue in Hawaii case." Retrieved from: http://www.nytimes.com/1982/03/08/us/value-of-2-dolphins-set-free-in-77-at-issue-in-hawaii-case.html
40. Midgley, M. (1985) "Persons and non-Persons." In Peter Singer (ed.), *In Defense of Animals*, New York: Basil Blackwell, pp. 52–62. Retrieved from: http://www.animal-rights-library.com/texts-m/midgley01.htm
41. Rector, JM (2014) *The Objectification Spectrum: Understanding and Transcending Our Diminishment and Dehumanization of Others*. Oxford University Press.
42. Zimbardo, P. (2008) *The Lucifer Effect: How Good People Turn Evil*, p. 223. Rider.
43. Jackson, R. (2005) *Writing the War on Terrorism: Language, Politics and Counter-Terrorism*, p. 60. Manchester University Press.

44. Zimbardo, *ibid*, p. 307.

Chapter 5: Thinking – Feeling

1. Thurman, J. (April 2013) "An unfinished woman: The desires of Margaret Fuller." *The New Yorker*. http://www.newyorker.com/magazine/2013/04/01/an-unfinished-woman

2. Aldridge, J. et al. (2011) "Parallels in the Beliefs and Works of Margaret Fuller and Carl Jung: Dreams, Literature, Spirituality, and Gender." *Sage Open*. Retrieved from: http://sgo.sagepub.com/content/1/1/2158244011410324

3. Fuller, M. (1843) "The Great Lawsuit. Man versus Men. Woman versus Women." *The Dial*. Retrieved April 2016 from http://transcendentalism-legacy.tamu.edu/authors/fuller/debate.html

4. Fuller, M. (1846) "A Dialogue: Critic, Poet." Retrieved December 2015 from http://www.poetryfoundation.org/learning/essay/238688

5. Damasio, A. (2006) *Descartes' Error: Emotion, Reason and the Human Brain*. Vintage Books.

6. Locke, J. (1689) *An Essay Concerning Human Understanding. Book IV, Chapter 19* ("Of Enthusiasm"). Retrieved 19th April 2016 from http://www.earlymoderntexts.com/assets/pdfs/locke1690book4.pdf

7. Kant, I. (1784) "What is Enlightenment?" Retrieved 19th April 2016 from http://www.columbia.edu/acis/ets/CCREAD/etscc/kant.html

8. Locke, J. (1689), *ibid*.

9. Popper, KR (1979) "The Bucket and the Searchlight: Two Theories of Knowledge." Appendix 1 in *Objective Knowledge: An Evolutionary Approach* (pp. 341–361). Oxford University Press.

10. Popper, KR (2001) *All Life is Problem Solving*, p. 142. Routledge.

11. Popper, KR, *ibid*, p. 15.

12. Procter, RN (2012) "The history of the discovery of the cigarette–lung cancer link: evidentiary traditions, corporate denial, global toll." *Tobacco Control, 21*, 87–91.

13. Minnameier, G. (2010) "The Logicality of Abduction, Deduction and Induction." In: M. Bergman, S. Paavola, AV Pietarinen, H. Rydenfelt (eds.), *Ideas in Action: Proceedings of the Applying Peirce Conference* (pp. 239–251). Helsinki: Nordic Pragmatism Network.

14. Lett, J. (1990) "A field guide to critical thinking." *Skeptical Inquirer*, 14.2.

15. Hackermüller, L. et al. (2003) "Wave nature of biomolecules and fluorofullerenes." *Physical Review Letters*, 91, 090408. http://journals.aps.org/prl/abstract/10.1103/PhysRevLet t.91.090408

16. Schlosshauer, M. et al. (2013) "A snapshot of foundational attitudes toward quantum mechanics." *Studies in History and Philosophy of Science Part B: Studies in History and Philosophy of Modern Physics*, 44, 222–230.

17. Bailin, S. (2002) "Critical thinking and science education." *Science & Education, 11*, 361–375.

18. Feynman, R. quoted in Wallace, BA (2006) *Contemplative Science: Where Buddhism and Neuroscience Converge*, p. 29. Columbia University Press.

19. Nickerson, R. (1998) "Confirmation bias: A ubiquitous phenomenon in many guises." *Review of General Psychology, 2*, 175–220.

20. *ibid.*

21. Ware, M. (2015) *The STM Report: An overview of scientific and scholarly journal publishing*. International Association of Scientific, Technical and Medical Publishers.

22. van Noorden, R. (2014) "Publishers withdraw more than 120 gibberish papers: Conference proceedings removed from subscription databases after scientist reveals that they were computer-generated." *Nature*, News, February 2014.

23. Bohannon, J. (2013) "Who's Afraid of Peer Review?" *Science, 342*, 60–65.

24. *Arguments for assisted dying:* (1) Confidential reports from doctors suggest that assisted dying is already widely practiced, for example by withdrawing treatment if the patient requests; (2) Certain terminal illnesses can lead to loss of dignity and constant pain for many years. If a person can voluntarily choose to avoid that suffering and pain, they should be allowed to; (3) Respecting an adult patient's autonomy and personal choice is central to a progressive, liberal society, and should be respected in matters of dying as well as living; (4) Comprehensive legislation can ensure that *only* terminally ill patients with extreme suffering and pain can access an assisted dying service.

   Arguments against assisted dying: (1) Assisted dying, if it became popular, could end up being seen as a cheaper alternative to investing in palliative care and hospice care; (2) A terminally ill person may well go through severe depression as they cope with their diagnosis, during which they may *temporarily* lose the will to live. Assisted dying services may be used by those who are depressed, before they have a chance to come to terms with their diagnosis; (3) If assisted dying is legal, then vulnerable patients could be forced or pressurized by relatives to take the option; (4) The terminal nature of an illness is sometimes diagnosed by mistake, and people may make an unexpected recovery; (5) Assisted dying legislation may help to promote an attitude in society that suffering should not be part of life, and/or that disability means that life is not worth living, leading to a 'slippery slope' of more and more euthanasia.

   For more details on these arguments, see Chapter 14 of my book *Development through Adulthood: An Integrative Sourcebook*, published by Palgrave Macmillan.

25. Brabazon, J. (2005) *Albert Schweitzer: A Biography*. Orbis

Books.

26. Vaillant, G. (2008) *Spiritual Evolution: A Scientific Defense of Faith*. New York: Broadway Books.

27. Gottlieb, RS (2012) *Spirituality: What It Is and Why It Matters*. Oxford University Press.

28. Freke, T. (2012) *The Mystery Experience: A Revolutionary Approach to Spiritual Awakening*. London: Watkins Publishing.

29. Aristotle, cited in Ryff, C. & Singer, BH (2008) "Know Thyself and Become What You Are: A Eudaimonic Approach to Psychological Well-Being." *Journal of Happiness Studies, 9*, 13–39.

30. H.H. Dalai Lama (1999) *The Art of Happiness: A handbook for living*. Part 1. Riverhead Books.

31. Rozin, P. & Royzman, EB (2001) "Negativity Bias, Negativity Dominance, and Contagion." *Personality and Social Psychology Review, 5*, 296–320.

32. Greenspan, M. (2004) *Healing Through the Dark Emotions: The wisdom of grief, fear, and despair*. Boston, MA: Shambhala.

33. Blake, W. (1863) *Auguries of Innocence*.

34. Becker, J. (2004) *Deep Listeners: Music, Emotion, and Trancing*, p. 66. Indiana University Press.

35. Roth, G. (1990) *Maps to Ecstasy*, p. 2. Aquarian Press.

36. Roth, G., *ibid*, p. 62 and p. 83.

37. Goodman, FD (1988) *Ecstasy, Ritual, and Alternate Reality: Religion in a Pluralistic World*. Indiana University Press.

38. Liszt, quoted in Blanning, T. (2012) *The Romantic Revolution*, p. 42. Modern Library.

39. Keats, J. (1817) "Letter to Benjamin Bailey." Retrieved from http://www.john-keats.com/briefe/221117.htm

40. Funk, MM (1998) *Thoughts Matter: The Practice of the Spiritual Life*. Continuum.

41. Otto, R. (1958) *The Idea of the Holy*. JW Harvey (trans.). London: Oxford University Press.

42. Burke, E. (1757) *A Philosophical Enquiry into the Origin of Our*

Ideas of the Sublime and Beautiful.

43. Kant, I., quoted in Morley, S. (2010) *The Sublime*, p. 16. MIT Press.

44. Mishra, V. (2010) "The gothic sublime," in Morley, S. (ed.), *The Sublime*, p. 155. Whitechapel Art Gallery.

45. Smith, A. (2008) *The Sublime in Crisis: Landscape Painting after Turner*. Retrieved 19th April 2016 from: http://www.tate.org.uk/art/research-publications/the-sublime/alison-smith-the-sublime-in-crisis-landscape-painting-after-turner-r1109220#f_1_6

46. Crowther, S. & Hall, J. (2015) "Spirituality and spiritual care in and around childbirth." *Women and Birth, 28*, 173–178. Retrieved April 2016 from http://www.sciencedirect.com/science/article/pii/S1871519215000037

47. Lintott, S. (2012) "The Sublimity of Gestating and Giving Birth: Toward a Feminist Conception of the Sublime," pp. 237–250. In S. Lintott & M. Sander-Staudt (eds.), *Philosophical Inquiries into Pregnancy, Childbirth, and Mothering*. London: Routledge.

48. Greenspan, M. (2004), *ibid.*

49. Ram Dass (1997) *Journey of Awakening: A Meditator's Guidebook*, p. 216. Bantam USA.

50. Cited in: Buehrens, JA & Forrester Church, F. (1989), *Our Chosen Faith: An Introduction to Unitarian Universalism*. Beacon Press.

51. Solomon, RC (2006) *Spirituality for the Skeptic: The Thoughtful Love of Life*. Oxford University Press.

52. Meitner, L. (1953) Lecture, Austrian UNESCO Commission (30 March 1953), in *Atomenergie und Frieden: Lise Meitner und Otto Hahn* (1953), 23–4. Trans. Ruth Sime, *Lise Meitner: A Life in Physics* (1996), 375.

53. Sagan, C. (1997) *The Demon-Haunted World: Science as a Candle in the Dark*, p. 45. Ballantine Books.

54. Myers, DG (2004) *Intuition: Its Powers and Perils*, p. 242. Yale

University Press.

55. Beveridge, WI (2004) *The Art of Scientific Investigation*. Blackburn Press.

56. Einstein, A. (1932) "Preface." In M. Planck, *Where is Science Going?* WW Norton & Company.

57. Bernard, C. cited in Beveridge, WI (2004) *The Art of Scientific Investigation*. Blackburn Press.

Chapter 6: Empirical – Transcendental

1. Armstrong, CJR (1975) *Evelyn Underhill: An introduction to her life and writings*. AR Mowbray & Co.

2. Williams, C. (ed.) (1944) *The Letters of Evelyn Underhill*. Longman.

3. Underhill, E. (1911) *Mysticism: A study of the nature and development of man's spiritual consciousness*, p. 26.

4. *ibid*, p. 19.

5. Lovejoy, AO (1936) *The Great Chain of Being: A Study of the History of an Idea*. Harvard University Press.

6. Smith, H. (1976) *Forgotten Truth: The Common Vision of the World's Religions*. HarperOne.

7. Cecil Powell's speech at the Nobel Banquet in Stockholm, December 10, 1950. http://www.nobelprize.org/nobel_prizes/physics/laureates/1950/powell-speech.html

8. Chapelle, F. (2005) *Wellsprings: A Natural History of Bottled Spring Waters*. Rutgers University Press.

9. Braile, LW & Braile, SJ (2001) *Plotting Earthquake Epicenters*. Retrieved from http://web.ics.purdue.edu/~braile/edumod/epiplot/epiplot.htm

10. Templeton, J. (2002) *Possibilities for Over One Hundredfold More Spiritual Information: The Humble Approach in Theology and Science*, p. 59. Templeton Foundation Press.

11. Swedenborg, E. (1758/2013) *Heaven and Hell*, Section 76. Jazzybee Verlag.

12. Mavromatis, A. (1987) *Hypnagogia: The Unique State of*

Consciousness Between Wakefulness and Sleep. Thyrsos Press.

13. Moody, R. (2001) *Life After Life.* Rider.
14. James, W. (1983) *The Varieties of Religious Experience,* p. 516. Penguin.
15. Steiner, R. (1994) *How to Know Higher Worlds: A Modern Path of Initiation.* Steiner Press.
16. Kelly, E., Crabtree, A. & Marshall, P. (2015) *Beyond Physicalism: Toward Reconciliation of Science and Spirituality.* Rowman and Littlefield.
17. Daniels, M. (2005) *Shadow, Self, Spirit: Essays in Transpersonal Psychology.* Imprint Academic.
18. Grof, S. (1991) *The Holotropic Mind: The Three Levels of Human Consciousness and How They Shape Our Lives,* p. 202. HarperCollins.
19. Grof, S. (1991), *ibid.*
20. Morley, C. (2013) *Dreams of Awakening: Lucid Dreaming and Mindfulness of Dream and Sleep.* Hay House.
21. Smith, H. (1976) *Forgotten Truth: The Common Vision of the World's Religions,* p. 75. HarperOne.
22. Grosman, L. et al. (2008) "A 12,000-year-old Shaman burial from the southern Levant (Israel)." *PNAS, 105,* 17665–17669.
23. Walsh, R. (1989) "What is a Shaman? Definition, Origin and Distribution." *The Journal of Transpersonal Psychology, Vol. 21.* No. I.
24. Ryan, RE (2001) *Shamanism and the Psychology of C.G. Jung: The Great Circle.* Vega Books; 1st Edition.
25. McKenna, T. (1998) *The Archaic Revival,* p. 250. Bravo Ltd.
26. Jung, CG (2009) *The Red Book: Liber Novus.* WW Norton & Company.
27. Huxley, A. (1994) *The Doors of Perception and Heaven and Hell,* pp. 12–13, 61. Flamingo.
28. Strassman, R. (2000) *DMT: The Spirit Molecule.* Park Street Press.
29. For example: Campos, DJ (2011) *The Shaman and Ayahuasca:*

Journeys to Sacred Realms. Studio City, CA: Divine Arts.

30. I wrote and published a personal testimony about a three-day ayahuasca workshop that I participated in during 2015. The workshop entailed three ceremonies, each of which lasted about six hours. Each was incredibly deep and profoundly challenging. From them, I gained enduring insights about the nature of how waking consciousness is one level of many, how the human intellect can only grasp a small portion of reality, and also the importance of love as an ultimate value for life and also as a source of order and coherence. For a pdf of the published report, please contact me and I will be happy to provide it. The reference is: Robinson, OC (2015) "Ayahuasca: A personal encounter with the miracle vine." *Network Review, Spring 2015*, pp. 24–25.

31. Frood, A. (2015) "Ayahuasca psychedelic tested for depression: Pilot study with shamanic brew hints at therapeutic potential." *Nature*, 6th April 2015, doi:10.1038/nature.2015.17252.

32. Freimoser, T. & Fountoglou, E. (2015) "Psychedelic Mind Online: Phenomenology & Effects of Transpersonal Experiences." *Breaking Convention: 3rd International Conference on Psychedelic Consciousness*. Retrieved from: https://vimeo.com/141606158

33. Thoricatha, W. (2016) "The Science of Ayahuasca: Inside the ICEERS Study at the Temple of the Way of Light." http://psychedelictimes.com/ayahuasca/the-science-ayahuasca-inside-iceers-study-temple-way-light/

34. Gonzalez, D. (2017) *Preliminary data on the long-term effects of ayahuasca in grief*. Presentation at Psychedelic Science 2017, Oakland, California.

35. Laszlo, E. (1994) "The 'Genius Hypothesis': Exploratory Concepts for a Scientific Understanding of Unusual Creativity." *Journal of Scientific Exploration, 8*, 257–264.

36. Romme, M. et al. (2009) *Living with Voices: 50 Stories of*

Recovery. PCCS Books.

37. Longden, E. (2013) *The voices in my head.* TED Talk. Retrieved from: https://www.ted.com/talks/eleanor_longden_the_voices_in_my_head

38. Belanti J., Perera, M., Jagadheesan, K. (2008) "Phenomenology of Near-death Experiences: A Cross-cultural Perspective." *Transcultural Psychiatry, 45*(1):121–33.

39. Van Lommel, P. (2011) *Consciousness Beyond Life: The Science of the Near-Death Experiences.* HarperOne.

40. *Do Children Have NDEs?* Retrieved from http://iands.org/images/stories/pdf_downloads/child.pdf

41. Sutherland, C. (1990) "Changes in religious beliefs, attitudes, and practices following near-death experiences: An Australian study." *Journal of Near-Death Studies, 9,* 21–31.

42. Van Lommel, P., Van Wees, R., Meyers, V. and Elfferich, I. (2001) "Near-death experience in survivors of cardiac arrest: a prospective study in the Netherlands." *Lancet, 358,* 2039–2045.

43. Lorimer, D. (1990) *Whole in One: The Near-death Experience and the Ethic of Interconnectedness.* Arkana.

44. Alcock, J. (1981) "Psychology and Near-Death Experiences." In Kendrick Frazier, *Paranormal Borderlands of Science.* Prometheus Books, pp. 153–169.

45. Greyson, B. (2010) "Implications of near-death experiences for a post materialist psychology." *Psychology of Religion and Spirituality, 2,* 37–45.

46. Agrillo, C. (2011) "Near-Death Experience: Out-of-Body and Out-of-Brain?" *Review of General Psychology, 15,* 1–10.

47. Parnia, S. et al. (2014) "AWARE – AWAreness during REsuscitation – A prospective study." *Resuscitation, 85,* 1799–1805.

48. Fontana, D. (2005) *Is There an Afterlife? A Comprehensive Overview of the Evidence.* London: O-Books.

49. Carr, B. (2015) "Hyperspatial Models of Matter and Mind."

In Kelly, EF, Crabtree, A. & Marshall, P., *Beyond Physicalism: Toward Reconciliation of Science and Spirituality*, pp. 227–274. Rowman & Littlefield.

50. Luke, DP (2010) "Rock Art or Rorschach: Is there More to Entoptics than Meets the Eye?" *Time & Mind: The Journal of Archaeology, Consciousness & Culture, 3*, 9–2.

51. Hameroff, S. & Chopra, D. (2012) "The 'Quantum Soul': A scientific hypothesis." In *Exploring Frontiers of the Mind-Brain Relationship*. Editors: A. Moreira-Almeida, F. Santana Santos. Springer.

Chapter 7: Mechanism – Purpose

1. Bacon, F. quoted in Broad, CD (1952), *Ethics and the History of Philosophy*, p. 122. Routledge and Kegan Paul.

2. Broad, CD (1926) *The Philosophy of Francis Bacon: An Address Delivered At Cambridge On The Occasion Of The Bacon Tercentenary*. Originally published in 1926 by Cambridge University Press. http://www.ditext.com/broad/bacon.html

3. Dodd, A. (1986) *Francis Bacon's Personal Life-Story, 2 vols.* London: Rider.

4. Crisinel, AS & Spence, C. (2010) "As bitter as a trombone: synesthetic correspondences in nonsynesthetes between tastes/flavors and musical notes." *Attention, Perception, & Psychophysics, 72*, 1994–2002.

5. Casadevall and Fang (2007) "Mechanistic Science." *Infection and Immunity*, vol. 77, p. 3518.

6. Engel, GS et al. (2007) "Evidence for wavelike energy transfer through quantum coherence in photosynthetic systems." *Nature, 446* (7137): 782–6.

7. Thagard, P. (2000) *How Scientists Explain Disease*. Princeton University Press.

8. Kepler, J. quoted in Burtt, EA (1932), *The Metaphysical Foundations of Modern Science, Second Edition*, p. 120. London: Kegan Paul.

9. de La Mettrie, J. (1748) *Man a Machine*. Excerpts taken from: http://faculty.humanities.uci.edu/bjbecker/RevoltingIdeas/week8d.html

10. Dawkins, R. (2006) *The Selfish Gene, 30th Anniversary Edition*, p. xxi. Oxford University Press.

11. Lodish, H. et al. (2000) *Molecular Cell Biology, 4th Edition*. Section 12.2 "The DNA Replication Machinery." New York: Freeman.

12. van Noorden, R. (2009) *Ribosome clinches the chemistry Nobel*. Retrieved from: http://www.nature.com/news/2009/091007/full/news.2009.981.html

13. Singleton, MR et al. (2004) "Crystal structure of RecBCD enzyme reveals a machine for processing DNA breaks." *Nature*. http://www.ncbi.nlm.nih.gov/pubmed/15538360

14. Alberts, B. (1998) "The Cell as a Collection of Protein Machines: Preparing the Next Generation of Molecular Biologists." *Cell*, 92, 291–4.

15. Epstein, R. (2016) "The Empty Brain: Your brain does not process information, retrieve knowledge or store memories. In short: your brain is not a computer." *Aeon magazine*. https://aeon.co/essays/your-brain-does-not-process-information-and-it-is-not-a-computer

16. Russell, B. (1902/1985) "A Free Man's Worship." Volume 12 of *The Collected Papers of Bertrand Russell, 1902–14*. Routledge.

17. Dawkins, R. (2001) *River Out of Eden: A Darwinian View of Life*. London: Phoenix.

18. Rees, M. (2001) *Just Six Numbers: The Deep Forces That Shape the Universe*. New York: Basic Books.

19. Larson, EJ & Witham, L. (1998) "Leading scientists still reject God." *Nature, 394*. http://www.nature.com/nature/journal/v394/n6691/full/394313a0.html

20. Moskowitz, C. (April 2016) "Are We Living in a Computer Simulation?" *Scientific American*.

21. Bollinger, RR, Barbas, AS, Bush, EL, Lin, SS & Parker, W.

(21 December 2007). "Biofilms in the large bowel suggest an apparent function of the human vermiform appendix." *Journal of Theoretical Biology*, 249, 826–831.

22. Dawkins, R. (2009) *The Purpose of Purpose*. Talk retrieved from: https://www.youtube.com/watch?v=mT4EWCRfdUg

23. Sheldrake, R. (2012) *The Science Delusion*, p. 38. London: Coronet.

24. Bergson, H. (1911) *Creative Evolution*. Translated by Arthur Mitchell. Henry Holt and Company.

25. Russel Wallace, A. (1914) *The World of Life: A Manifestation of Creative Power, Directive Mind and Ultimate Purpose*. London: Chapman and Hall.

26. Driesch, H. (1914) *The History & Theory of Vitalism* (CK Ogden, trans.). London: Macmillan.

27. Feuillet, L., Dufour, H. & Pelletier, J. (2007) "Brain of a white-collar worker." *The Lancet, 370*, no.9583, p. 262.

28. Sheldrake, R. (2009) *A New Science of Life, 3rd Edition*. Icon Books.

29. Reppert, SM et al. (2010) "Navigational mechanisms of migrating monarch butterflies." *Trends in Neuroscience, 33*(9), 399–406.

30. Sheldrake, R., *ibid*.

31. Sheldrake, R. (2011) *The Presence of the Past: Morphic Resonance and the Habits of Nature*. Allen and Unwin.

32. Boden, MA (1972) *Purposive Explanation in Psychology*. Boston, MA: Harvard University Press.

33. Carver, CS and Scheier, MF (1998) *On the Self-Regulation of Behavior*. Cambridge: Cambridge University Press.

34. Bandura, A. (2001) "Social Cognitive Theory: An Agentic Perspective." *Annual Review of Psychology*, 52, p. 3.

35. Dik, BJ et al. (2012) "Development and Validation of the Calling and Vocation Questionnaire (CVQ) and Brief Calling Scale (BCS)." *Journal of Career Assessment*, 20, 242–263.

36. Adler, F. (1905) *The Essentials of Spirituality*. New York:

James Pott & Co.

37. Schelling, F. (1809/2003) *Philosophical Inquiries into the Nature of Human Freedom*. Trans. J. Gutmann. Open Court Publishing Company.

38. Pinkard, T. (2002) *German Philosophy 1760–1860: The Legacy of Idealism*, p. 7. Cambridge: Cambridge University Press.

39. Wilber, K. (2000) *Sex, Ecology, Spirituality: The Spirit of Evolution*. Boston: Shambhala.

40. Hillman, J. (1997) *The Soul's Code: In Search of Character and Calling*. New York: Bantam.

41. Ware, B. (2012) *The Top Five Regrets of the Dying: A Life Transformed by the Dearly Departing*. Hay House.

42. Campbell, J. (2012) *The Hero with a Thousand Faces*. New World Library.

43. The Disney hero quest guide was called *A Practical Guide to The Hero with a Thousand Faces,* written by Christopher Vogler while working for Disney, and was later released as a 2007 book *The Writer's Journey: Mythic Structure for Writers* published by Michael Wiese Productions.

44. Hollis, J. (1993) *The Middle Passage: From Misery to Meaning in Midlife*. Inner City Books.

45. Heelas, P. cited in Mercadante, L. (2004) *Beliefs without Borders: Inside the Minds of the Spiritual but not Religious*. New York: Oxford University Press.

46. Nietzsche, F. (1888) *Twilight of the Idols*. (1990 Penguin Classics ed.)

47. Heelas, P. (1996) *The New Age Movement: Religion, Culture and Society in the Age of Postmodernity*. Wiley-Blackwell.

48. Leak, GK et al. (2007) "The relationship between spirituality, assessed through self-transcendent goal strivings, and positive psychological attributes." *Research in the Social Scientific Study of Religion, 18*, pp. 263–280. Brill Books.

49. *Visions: Notes of the Seminar given in 1930–1934 by CG Jung,* edited by Claire Douglas. Vol. 2. Princeton, NJ: Princeton

University Press, Bollingen Series XCIX, 1997, p. 923.

50. Jung, CG (1975) *Letters of C. G. Jung, Volume 2; Volumes 1951–1961*, p. 463. London: Routledge and Kegan Paul.

51. Gauquelin, M. et al. (1979) "Personality and position of the planets at birth: An empirical study." *British Journal of Clinical Psychology, 18*, 71–75.

52. Eysenck, HJ & Nias, DKB (1982) *Astrology: Science or Superstition?* London: Temple Smith, and New York: St Martin's.

53. Tarnas, R. (2008) *Cosmos and Psyche: Intimations of a New World View*. London: Penguin.

54. Toomey, D. (2007) *The New Time Travelers: A Journey to the Frontiers of Physics*. WW Norton & Company.

Chapter 8: Verbal – Ineffable

1. Wittgenstein, L. (2001) *Tractatus Logico-Philosophicus*. Routledge.

2. Gross, AG (1996) *The Rhetoric of Science*. Harvard University Press.

3. Bohr, N. in Petersen, A. (ed.), *Quantum Physics and the Philosophical Tradition*. Cambridge, MA: MIT Press, 1968.

4. Myers, G. (1990) *Writing Biology: Texts in the Social Construction of Scientific Knowledge*. Madison: University of Wisconsin Press.

5. Gross, AG (1996) *The Rhetoric of Science*. Harvard University Press.

6. Latour, B. (1987) *Science in Action*. Harvard University Press.

7. Dantzig, T. (1947) *Number: The Language of Science*. George Allen and Unwin Ltd.

8. Schneider, MS (2003) *A Beginner's Guide to Constructing the Universe: The Mathematical Archetypes of Nature, Art and Science*. Avon Books.

9. Dantzig, T., *ibid*.

10. Seife, C. (2000) *Zero: The Biography of a Dangerous Idea*.

Souvenir Press.

11. Galileo, 1623/1990, pp. 237–238, as cited in Godfrey-Smith, P. (2003), *Theory and Reality: an introduction to the philosophy of science*. University of Chicago Press.

12. Schneider, MS (2003) *A Beginner's Guide to Constructing the Universe: The Mathematical Archetypes of Nature, Art and Science*. Avon Books.

13. Schneider, *ibid*.

14. Ferguson, M. (1987) *The Aquarian Conspiracy: Personal and Social Transformation in Our Time*, p. 362. Jeremy P. Tarcher.

15. Maitland, S. (2009) *A Book of Silence: A journey in search of the pleasures and powers of silence*, p. 283. Granta Books.

16. Lane, B. (2007) *The Solace of Fierce Landscapes: Exploring Desert and Mountain Spirituality*. Oxford University Press.

17. Mother Teresa of Calcutta, cited in *Seeds of the Spirit: Wisdom of the Twentieth Century* (1995), edited by RH Bell & BL Battin. Westminster John Knox Press.

18. Lane, B. (2007), *ibid*.

19. Cited in Lane (2007), *ibid*, p. 62.

20. Spearing, AC (ed., trans.) (2001) *The Cloud of Unknowing and Other Works*. London: Penguin.

21. Quaker Quest (2004) *Twelve Quakers and God*, p. 16. Quaker Publications.

22. *ibid*, p. 18.

23. Barks, C. (2005) *Rumi: The Book of Love: Poems of Ecstasy and Longing*. HarperCollins Publishers.

24. Cleary, T. (trans.) (1993) *The Secret of the Golden Flower*. HarperSanFrancisco.

25. Tagore, Rabindranath (1991) "Stray Birds," in *Collected Poems and Plays*. Delhi: Macmillan India Ltd.

26. Krishnamurti, J. (2002) *Meditations*. Shambhala Publications Inc.

27. Jankélévitch, V. (2003) *Music and the Ineffable*. Translated by Carolyn Abbate. Princeton Press.

28. Taylor, B. (2016) *The Melody of Time: Music and Temporality in the Romantic Era.* Oxford University Press.
29. Cooper, M. (1970) *Beethoven: The Last Decade, 1817–1827.* London: Oxford University Press.
30. McGuire, EJ (no date) "Ralph Vaughan Williams: Spiritual Vagabond." *Society for the Arts in Religious and Theological Studies, Vol. 27.* http://www.societyarts.org/arts-journal/online-edition/136-online-edition-vol-27-no-1/336-ralph-vaughan-williams-spiritual-vagabond
31. Quoted in: Blanning, T. (2008) *The Triumph of Music*, p. 115. London: Penguin Books.
32. Lester, P. (2014) "Mike Oldfield: 'We wouldn't have had Tubular Bells without drugs'." Retrieved from https://www.theguardian.com/music/2014/mar/20/mike-oldfield-interview-tubular-bells-drugs?CMP=Share_iOSApp_Other
33. From a letter by Mondrian to Steiner, in: Blotkamp, C. (2001) *Mondrian: The Art of Destruction*, p. 182. London: Reaktion Books Ltd.
34. Neumeyer, A. (1964) *The Search for Meaning in Modern Art*, p. 89. Prentice-Hall Inc.
35. Armstrong, K. (2004) *The Battle for God: Fundamentalism in Judaism, Christianity and Islam.* HarperCollins.
36. Huxley, A. (1957) *The Perennial Philosophy.* Chatto & Windus.
37. Teasdale, W. (2001) *The Mystic Heart: Discovering a Universal Spirituality in the World's Religions.* New World Library.

Chapter 9: Explanation – Contemplation

1. James, W. (1983) *The Varieties of Religious Experience*, p. 501. Penguin Classics.
2. Windelband, W. & Oakes, G. (1894/1980) "History and natural science." *History and Theory*, 19(2),165–168.
3. Pinker, S. (2013) "Science Is Not Your Enemy." *New Republic.* http://www.newrepublic.com/article/114127/science-not-enemy-humanities#

4. Feynman, R. (2011) *Six Easy Pieces: The Fundamentals of Physics Explained*, p. 69. London: Penguin.
5. Estelle, M. (1996) "Plant tropisms: The ins and outs of auxin." *Current Biology, 6*, 1589–1591.
6. Hempel, CG (1948) "Studies in the logic of explanation." *Philosophy of Science, 15*, 135–175. Retrieved from http://www.sfu.ca/~jillmc/Hempel%20and%20Oppenheim.pdf
7. Explanation of candle-glass experiment: Four mechanisms are necessary to explain the effect. Firstly the burning flame leads to the loss of twice the amount of oxygen as is replaced by carbon dioxide, leading to a decrease in air pressure within the candle. The second mechanism is temperature; when the candle goes out the gas cools and contracts as atoms move less and come closer together, leading to a drop in pressure. Third is the condensation effect; combustion produces water as a gas, which condenses into liquid when the water cools, leading to further loss of pressure within the glass. Fourthly is the mechanism of pressure itself; low pressure creates suction due to molecules moving in to fill what would otherwise be a vacuum. Interestingly, there is still a lack of consensus on why the water rises so rapidly after the candle goes out, rather than gradually as the air cools.
8. Feynman, R. (2011), *ibid.*
9. Penrose, R. (2005) *The Road to Reality: A Complete Guide to the Laws of the Universe*, pp. 13 and 22. Vintage.
10. Bacon, F. (1620) *Novum Organum XVII*. Bacon writes: "For when I speak of forms, I mean nothing more than those laws and determinations of absolute actuality which govern and constitute any simple nature, as heat, light, weight, in every kind of matter and subject that is susceptible of them."
11. Sagan, C. & Head, T. (2006) *Conversations with Carl Sagan (Literary Conversations Series)*. University Press of Mississippi.
12. Emerson, RW (1982) *Nature and Other Essays*, p. 47. Penguin.

13. Blake, W. & Jonson, W. (2012) *Auguries of Innocence and Other Lyric Poems*. CreateSpace Independent Publishing Platform.
14. Bortoft, H. (1996) *The Wholeness of Nature: Goethe's Way of Science*. Edinburgh: Floris Books.
15. *ibid*, p. 80.
16. Schopenhauer, A. (1995) *The World as Will and Idea*, p. 239. Phoenix Everyman.
17. Tolle, E. (1999) *The Power of Now: A Guide to Spiritual Enlightenment*, p. 63. Namaste Publishing.
18. Tolle, E. (2003) *Stillness Speaks: Whispers of Now*, p. 83. Hodder Mobius.
19. Huxley, A. (1994) *The Doors of Perception and Heaven and Hell*, pp. 7–8. Flamingo.
20. Stambuk, D. (2014) *Haiku Consciousness*. IAFOR Publications.
21. Kohak, E. (1980) *Idea and Experience: Edmund Husserl's Project of Phenomenology in "Ideas I"*, p. 60. University of Chicago Press.
22. Robinson, OC (2007) *Developmental Crisis in Early Adulthood: A composite qualitative analysis*. Doctoral Dissertation, Birkbeck College.
23. Reason, P. and Heron, J. (1995) "Co-operative Inquiry." In R. Harre, J. Smith & L. Van Langenhove (eds.), *Rethinking Methods in Psychology* (pp. 122–142). London: Sage Publications.
24. *ibid*.
25. Heron, J. (1992) *Feeling and Personhood: Psychology in Another Key*. London: Sage Publications.
26. Harner, M. (1990) *The Way of the Shaman, Third Edition*. HarperCollins, p. xii.
27. James, W. (1890) *The Principles of Psychology*. New York: Henry Holt and Company.
28. *British Religion in Numbers*. Based at University of Manchester. www.brin.ac.uk
29. Saucier, G. & Skrzypinska, K. (2006) "Spiritual But Not

Religious? Evidence for Two Independent Dispositions." *Journal of Personality, 74,* 1257–1292.

30. Robinson, OC (2013) *Development through Adulthood: An Integrative Sourcebook.* Basingstoke: Palgrave Macmillan.

31. Alister Hardy Religious Experience Research Centre, University of Wales Trinity Saint David. http://www.uwtsd. ac.uk/library/alister-hardy-religious-experience-research-centre

32. Fenwick, P. & Fenwick, E. (2008) *The Art of Dying.* London: Continuum.

33. Wade, J. (2008) *Transcendent Sex.* Simon and Schuster.

34. Heelas, P. & Woodhead, L. (2005) *The Spiritual Revolution: Why Religion is Giving Way to Spirituality.* Blackwell Publishing.

Chapter 10: MODI and the Wisdom of the Whole

1. Jung, CG (1995) *Memories, Dreams, Reflections,* p. 367. Flamingo.

2. Durant, W. (1957) "What is wisdom?" *Wisdom, II,* No. 8, 25–26. http://will-durant.com/wisdom.htm

3. Grossman, I. et al. (2016) "A Heart and a Mind: Self-distancing Facilitates the Association Between Heart Rate Variability, and Wise Reasoning." *Frontiers in Behavioral Neuroscience, 10,* Article 68.

4. Goodall, J. (2004) *Reason for Hope: A Spiritual Journey.* Grand Central Publishing.

5. Xinyan, J. (2013) "Chinese Dialectical Thinking – The Yin Yang Model." *Philosophy Compass, 8/5,* 438–446.

6. Jung, CG (1963) *Mysterium Coniunctionis: An Inquiry into the Separation and Synthesis of Psychic Opposites in Alchemy.* Routledge.

7. Hauck, D. (no date) *The AZoth Ritual.* http://azothalchemy. org/azoth_ritual.htm

8. McGilchrist, I. (2012) *The Master and His Emissary: The Divided*

Brain and the Making of the Western World. Yale University Press.

9. Muckli, L. et al. (2009) "Bilateral visual field maps in a patient with only one hemisphere." *PNAS, 106,* 13034–13039.

10. MacNeilage, PF, Rogers, LJ & Vallortigara, G. (2009) "Evolutionary Origins of Your Right and Left Brain. *Scientific American.* www.scientificamerican.com/article/ evolutionary-origins-of-your-right-and-left-brain

11. McGilchrist, I. (2011) "Can the divided brain tell us anything about the ultimate nature of reality?" Royal College of Psychiatrists. Retrieved from: https//www.rcpsych.ac.uk/ pdf/Iain%20McGilchrist%20Can%20the%20divided%20 brain%20tell%20us%20anything%20about%20the%20 ultimate%20nature%20of%20reality.pdf

12. Jung, CG & Pauli, W. (1952) *The Interpretation of Nature and the Psyche.* Ishi Press.

13. Planck, M. (1931): "I regard consciousness as fundamental. I regard matter as derivative from consciousness. We cannot get behind consciousness. Everything that we talk about, everything that we regard as existing, postulates consciousness." Quoted in *The Observer,* 25 January 1931.

14. Capra, F. (1991) *The Tao of Physics: An Exploration of the Parallels between Modern Physics and Eastern Mysticism.* Flamingo.

Epilogue: The Interconnected Age

1. Lagarde, C. (2013) "The Interconnected Global Economy: Challenges and Opportunities for the United States – and the World." International Monetary Fund. https://www.imf. org/en/News/Articles/2015/09/28/04/53/sp091913

2. Royal Society for Public Health (2017) "Status of Mind: Social media and young people's mental health and wellbeing." Retrieved from: https://www.rsph.org.uk/our-work/policy/ social-media-and-young-people-s-mental-health-and-

wellbeing.html

3. Baxter, AJ et al. (2014) "Challenging the myth of an 'epidemic' of common mental disorders: Trends in the global prevalence of anxiety and depression between 1990 and 2010." *Depression and Anxiety, 31,* 506–516.

4. Moss, R. (2017) "Majority of People Face Mental Health Problems, With 'Two In Three Adults Affected'." Retrieved from: http://www.huffingtonpost.co.uk/entry/mental-health-statistics-two-in-three-adults-will-face-a-problem_uk_5909ddeee4b02655f842c0e7

5. Robinson OC. (2010) "Modernity and the trans-modern shift." In *A New Renaissance: Transforming science, spirit and society* (Lorimer, D. & Robinson, OC, eds.). Edinburgh: Floris Books.

Index

78-79, 85
phenomenology 197, 211-212, 214
photons 29, 98-101, 144, 171, 182, 199
photosynthesis 150
physicalism 225-226
placebo 70
Planck, Max 98, 227
plate techtonics 122
Plato 21, 50, 133, 171, 184, 201-203, 227
pluralism 8, 32-33, 115
poetry 27, 74, 79, 81, 111, 190, 208-209
Polkinghorne, John 155
Pope, Alexander 119
Popper, Karl 94-96
postmodernism 32-33, 232
prayer 8, 10, 32, 52, 67, 77, 85
Priestley, Joseph 25-26
primary and secondary qualities 27-29, 40
problem-solving 75, 94-96
psychedelics 131, 136-138
psychoanalysis 36-37, 49, 62
psychology, science of 1
psychometrics 47-48, 214
psychotherapy 10, 58, 62, 109
Pythagoras 171, 179-183
Quakers 9, 23-24, 50, 187
qualitative methods 47-48, 211, 214-215
quantum entanglement 130,

144, 228
quantum physics / quantum mechanics 38, 64, 98, 100, 143, 147, 150, 201, 230
radio waves 45, 124, 126
Ram Dass 114
Reason, Peter 212-213
reductionism 149-150, 224
Rees, Martin 154
relativity theory 28-29, 33, 39, 143, 171, 172
religion, definition of 7-9
Renaissance 11, 135
Restall Orr, Emma 85
rhetoric 174-176
romantic music 27
romantic revolution 24, 26
Romanticism 26-28, 79, 111, 204-205
Rosetta space probe 147
Rosicrucianism 24, 146
Roth, Gabrielle 109-110
Royal Society, the 22, 121, 145, 154
Rumi 74, 188
Russel Wallace, Alfred 41, 157
Russell, Bertrand 153-154, 173
Russell, Peter 54
sacred geometry 182-184, 219
Sagan, Carl 42, 116, 226
Schelling, Friedrich Wilhelm Joseph 164-166, 227
Schopenhauer, Arthur 64, 82, 205, 227

BOOKS

SPIRITUALITY

O is a symbol of the world, of oneness and unity; this eye represents knowledge and insight. We publish titles on general spirituality and living a spiritual life. We aim to inform and help you on your own journey in this life. If you have enjoyed this book, why not tell other readers by posting a review on your preferred book site? Recent bestsellers from O-Books are:

Heart of Tantric Sex
Diana Richardson
Revealing Eastern secrets of deep love and intimacy to Western couples.
Paperback: 978-1-90381-637-0 ebook: 978-1-84694-637-0

Crystal Prescriptions
The A-Z guide to over 1,200 symptoms and their healing crystals
Judy Hall
The first in the popular series of six books, this handy little guide is packed as tight as a pill-bottle with crystal remedies for ailments.
Paperback: 978-1-90504-740-6 ebook: 978-1-84694-629-5

Take Me To Truth
Undoing the Ego
Nouk Sanchez, Tomas Vieira
The best-selling step-by-step book on shedding the Ego, using
the teachings of A Course In Miracles.
Paperback: 978-1-84694-050-7 ebook: 978-1-84694-654-7

The 7 Myths about Love...Actually!
The journey from your HEAD to the HEART of your SOUL
Mike George
Smashes all the myths about LOVE.
Paperback: 978-1-84694-288-4 ebook: 978-1-84694-682-0

The Holy Spirit's Interpretation of the New Testament
A course in Understanding and Acceptance
Regina Dawn Akers
Following on from the strength of *A Course In Miracles*, NTI
teaches us how to experience the love and oneness of God.
Paperback: 978-1-84694-085-9 ebook: 978-1-78099-083-5

The Message of A Course In Miracles
A translation of the text in plain language
Elizabeth A. Cronkhite
A translation of *A Course in Miracles* into plain, everyday
language for anyone seeking inner peace. The companion
volume, *Practicing A Course In Miracles*, offers practical lessons
and mentoring.
Paperback: 978-1-84694-319-5 ebook: 978-1-84694-642-4

Thinker's Guide to God
Peter Vardy
An introduction to key issues in the philosophy of religion.
Paperback: 978-1-90381-622-6

Your Simple Path
Find happiness in every step
Ian Tucker
A guide to helping us reconnect with what is really important
in our lives.
Paperback: 978-1-78279-349-6 ebook: 978-1-78279-348-9

365 Days of Wisdom
Daily Messages To Inspire You Through The Year
Dadi Janki
Daily messages which cool the mind, warm the heart and guide
you along your journey.
Paperback: 978-1-84694-863-3 ebook: 978-1-84694-864-0

Body of Wisdom
Women's Spiritual Power and How it Serves
Hilary Hart
Bringing together the dreams and experiences of women across
the world with today's most visionary spiritual teachers.
Paperback: 978-1-78099-696-7 ebook: 978-1-78099-695-0

Dying to Be Free
From Enforced Secrecy to Near Death to True Transformation
Hannah Robinson
After an unexpected accident and near-death experience,
Hannah Robinson found herself radically transforming her life,
while a remarkable new insight altered her relationship with
her father; a practising Catholic priest.
Paperback: 978-1-78535-254-6 ebook: 978-1-78535-255-3

The Ecology of the Soul
A Manual of Peace, Power and Personal Growth for Real People
in the Real World
Aidan Walker
Balance your own inner Ecology of the Soul to regain your
natural state of peace, power and wellbeing.
Paperback: 978-1-78279-850-7 ebook: 978-1-78279-849-1

Not I, Not other than I
The Life and Teachings of Russel Williams
Steve Taylor, Russel Williams
The miraculous life and inspiring teachings of one of the
World's greatest living Sages.
Paperback: 978-1-78279-729-6 ebook: 978-1-78279-728-9

On the Other Side of Love
A Woman's Unconventional Journey Towards Wisdom
Muriel Maufroy
When life has lost all meaning, what do you do?
Paperback: 978-1-78535-281-2 ebook: 978-1-78535-282-9

Readers of ebooks can buy or view any of these bestsellers by
clicking on the live link in the title. Most titles are published
in paperback and as an ebook. Paperbacks are available in
traditional bookshops. Both print and ebook formats are
available online.

Find more titles and sign up to our readers' newsletter at
http://www.johnhuntpublishing.com/mind-body-spirit

Follow us on Facebook at https://www.facebook.com/OBooks/
and Twitter at https://twitter.com/obooks